Role of Surface Modification on Bacterial Adhesion of Bio-Implant Materials

Role of Surface Modification on Bacterial Adhesion of Bio-Implant Materials

Machining, Characterization, and Applications

Santhosh Kumar S
Somashekhar S. Hiremath

CRC Press
Taylor & Francis Group
Boca Raton London New York

CRC Press is an imprint of the
Taylor & Francis Group, an **informa** business

First edition published 2021
by CRC Press
6000 Broken Sound Parkway NW, Suite 300, Boca Raton, FL 33487-2742

and by CRC Press
2 Park Square, Milton Park, Abingdon, Oxon, OX14 4RN

First issued in paperback 2022

CRC Press is an imprint of Taylor & Francis Group, an Informa business

No claim to original U.S. Government works

ISBN 13: 978-0-367-53132-4 (pbk)
ISBN 13: 978-0-367-89458-0 (hbk)
ISBN 13: 978-1-003-02140-7 (ebk)

DOI: 10.1201/9781003021407

Visit the Taylor & Francis Web site at
http://www.taylorandfrancis.com

and the CRC Press Web site at
http://www.crcpress.com

Library of Congress Cataloging-in-Publication Data

Names: Hiremath, Somashekhar S., author. | Kumar, S. (Santosh), 1953- author.
Title: The role of surface modification on bacterial adhesion of bio-implant materials : machining, characterization, and applications / Somashekhar S. Hiremath and Santhosh Kumar.
Description: First edition. | Boca Raton : CRC Press, 2020. | Includes bibliographical references and index.
Identifiers: LCCN 2020017205 | ISBN 9780367894580 (hardback) | ISBN 9781003021407 (ebook)
Subjects: LCSH: Biomedical materials. | Biomedical materials–Biocompatibility. | Biomedical materials–Surfaces. | Implants, Artificial–Materials.
Classification: LCC R857.M3 H57 2020 | DDC 610.285–dc23
LC record available at https://lccn.loc.gov/2020017205

Contents

Preface

Orthopaedic joint implants such as hip, knee, shoulder, ankle, and elbow prosthesis are used for hard tissue replacement. These are load-bearing joints subjected to high levels of mechanical stress, fatigue, and wear in the normal daily activities of humans. The artificial implants used for these applications should possess structural integrity as well as surface compatibility with the surrounding biological environment for prolonged survivability without complications. Despite the adoption of advanced technology in manufacturing new implants and in surgical and medical management procedures, there are still a large number of implants that are subject to infection. The major causes for implants failure are a fracture, prosthetic dislocation, loosening, excessive wear rate at mating surfaces and its associated debris (although this cause has been much reduced by the use of cross-linked polyethylene), and pre-surgical contamination/infection (i.e. bacterial adhesion). In this book, a focus has been made on the implant failure due to bacterial infections, which is critically important from a clinical perspective when surfaces are made with a finishing process. Bacterial infections on implants are usually initiated through adhesion of bacteria to the implant surface by means of physiochemical interaction between the implant surface and bacteria, that is van der waals forces, electrostatic forces, and gravitational forces. The reversible bacterial adhesion is followed by colonization and formation of biofilm that forms a layer of bacteria binding irreversibly to the implant surface, which is later difficult to remove from implants. The bacterial adhesion on the implant surface depends on the (1) properties of the implant material such as surface topography, surface roughness, surface chemistry, and surface energy; (2) properties of the bacteria such as surface charge, surface hydrophobicity hydrophilicity, and appendages; and (3) properties of the surrounding environment such as type of antibiotics and associated flow condition, temperature, pH value, a period of exposure, chemical treatment, and bacterial concentration. In particular, surface roughness and topography are the most influencing parameters because studies show that surface roughness increase the surface area of the material and also the depression, grooves, pits, scratches, and crevices in the rough surfaces influences the bacterial adhesion and acts as favourable sites for colonization and biofilm formation.

The near-net-shape of the bio-implants with the required surface finish is achieved by various finishing steps after the primary machining process. After the pre-fabrication, implants are subjected to the grinding process at the end. Final finishing of these surfaces is performed using polishing integrated with multi-axis CNC machines, vibratory abrasive polishing, free abrasive polishing, fixed abrasive polishing, belt polishing, and so on, which have all been comprehensively used in the past few decades to remove the finer irregularities and these techniques along with some of the advanced surface modification techniques are explained in this book.

Most of the implants have a complex and partly freeform surface that raises the difficulty in finishing at one step, and also one should have the knowledge of frequently changing contact conditions while performing finishing operation. There is a need for a technique that can be used to finish the complex surfaces of orthopaedic implants without changing its contact condition and without disturbing the dimensional accuracy. Abrasive flow finishing is one such process, in which flexible abrasive media is used to finish the complex surfaces. Previous studies have demonstrated that this process is capable of finishing complex internal and external surfaces and also freeform surfaces with controlled surface topography without disturbing the dimensional accuracy. Although there have been studies on the finishing of freeform surfaces using this process, understanding the biocompatibility of surfaces finished with this process is still far from the complete study.

In the present research work, to overcome the challenges associated with conventional finishing processes to finish complex internal and external surfaces along with free form surfaces, a unidirectional abrasive flow finishing (UAFF) process is developed in which abrasive media flows only in one direction to achieve the uniform lay patterns on the work surface in one direction with the required finish, which finds wide application in lubrication, biomedical, and surface engineering. The experimental setup mainly consists of the hydraulic power pack, a hydraulic cylinder, an abrasive media cylinder, direction control valves, and pressure gauges.

In the present work, a viscoelastic polymer-based abrasive media is developed using a two-roll mill process. The selected constituents are viscoelastic polymer: silicone rubber (38%), plasticizer: silicone oil (12%) reinforced with suitable abrasive particles: SiC (50%). The developed abrasive media is subjected to different characterization studies such as the morphology using a high-resolution scanning electron microscope (HRSEM) – Inspect F50, thermal properties using thermo gravimetric analysis (TGA) – SDT Q600 instrument, and differential scanning calorimetry (DSC) – Model Q2000 V24 instrument. The tensile property of the abrasive media is measured using the Universal testing machine –ZwickRoell. Further, rheological properties such as viscosity, shear stress, storage modulus, loss modulus, loss tangent, and complex viscosity are ascertained using the Anton Paar Physica MCR 301 rheometer instrument. The rheological properties are studied at different temperatures: 25°C, 35°C, 45°C, and 55°C to study the behaviour of abrasive media under different temperature condition.

The experimentation has been carried out at various levels to investigate the performance of the developed unidirectional abrasive flow finishing setup. Machinability study has been carried out on different engineering materials having different hardness values such as aluminium, brass, copper, and mild steel machined with varying number of cycles. The objective of this experiment is to finish the biomaterials for different cycles and to study the effect of the UAFF process on the surface roughness, surface morphology, bearing area curve (BAC), and wettability of the biomaterial surfaces. The selected process parameters are number of cycles: 3, 6, 9, and pressure: 40 bar, 50 bar, and 60 bar. The obtained surface roughness and BAC are measured using optical profilometer Wyko NT1100 instrument, and morphology of the surface is studied using an HRSEM image. Further, the wettability of the finished surface is studied by measuring the contact angle (θ) for three different liquids – water, formamide, and diiodomethane by sessile drop technique using contact angle Goniometer – GBX-Digidrop MCAT instrument. The measured contact angles are used to ascertain the surface free energy using the van-Oss-Chaudhury-Good equation. Further, response surface methodology (RSM) is used to optimize the input parameters such as a number of cycles and pressure to obtain the desired output responses, namely, average surface roughness (R_a), and material removed (MR).

Bacterial adhesion study has been carried out on machined biomaterials to investigate the influence of process parameters on surface roughness and bacterial adhesion. The developed setup has been used to finish the biomaterials – stainless steel (SS316L) and titanium alloy (Ti-6Al-4V) with two different abrasive media (220 mesh and 400 mesh) and a varying number of cycles (3 and 6). The purpose of this study is to evaluate the effect of this finishing process on enhancing surface characteristics such as surface roughness, surface morphology, and its role in initial bacterial adhesion. Optical profilometer and an HRSEM are used to examine the surface roughness and surface morphology of the finished surfaces. Further, for the bacterial adhesion study, both *Escherichia coli* (*E. coli*) and *Staphylococcus aureus* (*S. aureus*) are selected because these bacterial strains are commonly found on orthopaedic implant-related infections.

The procedure followed during the experiments and experimental findings is detailed in the book under various chapters with different headings along with an introduction chapter, which covers the different surface modification techniques.

Structure of the Book

Chapter 1 covers introduction to bio-implants, classifications of various implants, different materials used to manufacture implants, the reason for implant failures, failure due to bacterial adhesion and biofilm formation, and a broad literature survey.

A wide variety of established methods developed for implant surface modification such as mechanical, chemical, and thermal processes are summarized in **Chapter 2**. Further, this chapter mainly focuses on implant surface roughness and topographical features that influence implant failures. After obtaining a panoramic view on the topic through the literature survey, the objective and scope of the book are carried out, followed by the motivation.

Chapter 3 introduces the readers to different types of finishing processes, the importance of finishing processes, surface topography, and surface morphology. Different variants of abrasive flow machining, material removal mechanism, and process parameters are discussed in detail. A literature survey on various aspects of the abrasive flow machining, motivation, objectives, and methodology of the research work is presented in this chapter.

Chapter 4 contains the development and fabrication details of the experimental setup carried out at different stages. It also includes the information on materials and various fixtures developed in the present work to carry out the experimentation. This chapter also consists of a selection of materials and preparation steps followed in the development of abrasive media. This section also highlights the different characterization studies used to study the properties of the abrasive media and results.

Chapter 5 presents the results and analysis of the experimentation carried out in the present research work. Machinability study on different work materials has been carried out. Further, surface roughness and morphology of the abrasive flow finished work surfaces on wettability, and bacterial adhesion is presented at the end of the chapter.

Chapter 6 includes conclusions drawn based on the experimental studies carried out in the present research work, and the scope for future work is highlighted. The contribution of the work is also listed in this chapter.

Abbreviations

AA	Aluminium alloy
AFF	Abrasive flow finishing
AFM	Abrasive flow machining
AISI	American Iron and Steel Institute
Al	Aluminium
Al_2O_3	Aluminium oxide
ANN	Artificial Neural Network
ANOVA	Analysis of variance
AOA NJRR	Australian Orthopedic Association National Joint Replacement Registry
ASTM	American Society for Testing and Materials
ATCC	American Type Culture Collection
B_4C	Boron carbide
BAC	Bearing area curve
CCD	Central composite design
C-D	Convergent-Divergent
CFAAFM	Centrifugal force assisted abrasive flow machining
CFD	Computational fluid dynamics
CLUAFF	Closed loop unidirectional abrasive flow finishing
CNC	Computer numerical control
Cu	Copper
DBG-AFF	Drill bit guided-abrasive flow finishing
DCV	Directional control valve
DDS	Data dependent systems
DI	Deionized water
DSC	Differential scanning calorimetry
E. coli	*Escherichia coli*
ECAFM	Electro-chemical aided abrasive flow machining
EDAX	Energy dispersive X-ray analysis

EDM	Electrical discharge machining
GA	Genetic algorithms
GMDH	Group method of data handling
GP	Gel point
GRA	Grey relational analysis
HB	Hardness Brinell
HRSEM	High resolution scanning electron microscope
ID	Internal diameter
LB	Lysogeny broth
MAFM	Magneto abrasive flow machining
MEMS	Micro-electro-mechanical-systems
MR	Material removed
MRAFF	Magneto rheological abrasive flow finishing
MRR	Material removal rate
MS	Mild steel
MWD	Molecular weight distribution
OD	Outer diameter
PII	Prosthetic implant infection
R-AFF	Rotational-abrasive flow finishing
R-MRAFF	Rotational magneto rheological abrasive flow finishing
RMS	Root mean square
RSM	Response surface methodology
S. aureus	*Staphylococcus aureus*
S/N	Signal-to-noise ratio
SEM	Scanning electron microscope
SiC	Silicon carbide
SR	Silicone rubber
SS	Stainless steel
TGA	Thermo gravimetric analysis
Ti	Titanium
UAAFM	Ultrasonic assisted abrasive flow machining
UAFF	Unidirectional abrasive flow finishing
USA	United States of America
UV	Ultraviolate
V	Vanadium

Notations

$\dot{\gamma}$	Shear rate, 1/s
γ^-	Base component, mN/m
$\theta^\perp$	Contact angle – Perpendicular direction, deg
$\theta^\parallel$	Contact angle – Parallel direction, deg
γ^+	Acid component, mN/m
γ^{LW}	Lifshitz–van der Waals or dispersive component, mN/m
γ^{TOT}	Total surface tension of the liquid, mN/m
ω	Angular velocity, rad/s
δ	Oscillation phase angle, deg
τ	Shear stress, Pa
η	Viscosity, Pa.s
$\Delta\theta$	Change in contact angle, deg
η'	Dynamic viscosity (In-phase), Pa.s
η''	Dynamic viscosity (Out-of-phase), Pa.s
η^*	Complex viscosity, Pa.s
ϕ_0	Angular amplitude of oscillation
$\rho_a,\ \rho_p,\ \rho_o$	Density of abrasive particles, density of polymer and density of oil, kg/m^3
θ_D	Contact angle – Diiodmethane, deg
θ_F	Contact angle – Formamide, deg
ρ_m	Density of abrasive media, kg/m^3
ΔR_a	Change in average surface roughness, μm
η_t	Total efficiency
θ_W	Contact angle – Water, deg
$A1$	Area of peak, μm^2
$A2$	Area of valley, μm^2
A_a	Area of abrasive cylinder, m^2
A_f	Area of hydraulic cylinder piston end (forward direction), m

A_r	Area of hydraulic cylinder rod end (return direction), m
D	Diiodomethane
d_a	Diameter of abrasive media cylinder, m
d_w	Maximum internal diameter of the workpiece, m
F	Formamide
f	Frequency, Hz
F_a	Axial force, N
F_r	Radial force, N
G'	Storage modulus, Pa
G''	Loss modulus, Pa
h	Thickness of the sample or gap between two plates, mm
l_s	Stroke length, m
M	Number of peaks in the profile
M	Torque, N m
M_0	Torque amplitude, Nm
M_{r1}	Peak material component
M_{r2}	Valley material component
P	Power, kW
P_c	Peak count
p_h	Pressure at hydraulic cylinder, bar
P_{max}	Maximum operating pressure, bar
Q_{max}	Maximum media flow rate, kg/m^3
r	Radius of the plate, mm
R_a	Average surface roughness, μm
R_k	Core roughness depth, μm
R_{max}	Maximum roughness depth, μm
R_p	Maximum height of peaks, μm
R_{pk}	Reduced peak height, μm
R_q	Root-mean-square roughness, μm
R_t	Maximum height of the profile, μm
R_v	Maximum depth of valleys, μm
R_{vk}	Reduced valley depth, μm
R_z	Ten-point height, μm
S	Mean spacing of adjacent local peaks, μm
t	Time, s
V	Volume of working cylinder, m^3
v_a	Volume of the abrasive cylinder, m^3
v_{max}	Maximum media flow velocity, m/s
W	Water

w_a, w_p, w_o Weight percentage of abrasive particles, wt% of polymer and wt% of oil

W_c Weight of the constituent, kg

w_c Weight percentage of each constituent

w_m Total weight of abrasive media, kg

Δ_q RMS slope of the profile

λ_q RMS wave length, μm

Introduction to Bio-Implants

1.1 INTRODUCTION

Bio-implants are artificial medical devices used to replace the missing biological system and the damaged biological system, and also to support or enhance the existing biological structure. These implants can be of many types, such as load-bearing implants used in orthopaedic applications, dental implants used in restoring the functionality and appearance of natural teeth, implants used in cardiovascular system, and many more. Metallic biomaterials are widely used among other materials, especially in load-bearing implants and internal fixation devices, due to their superior characteristics such as high tensile strength, high yield strength, resistance to fatigue loading, resistance to creep, high corrosion resistance, and biocompatibility. Most popular metallic biomaterials in use today are stainless steel, cobalt-chromium alloys, titanium, and titanium alloys. These materials are used in cardiovascular, orthopaedic, dentistry, craniofacial, and otorhinology implants. The implants are usually manufactured using milling, casting, forging, compression moulding, powder metallurgy, and rapid prototype techniques; further features are generated using drilling, electrical discharge machining (EDM), electron beam machining (EBM), laser beam machining (LBM), and ultrasonic machining (USM).

According to statistics in the United States, around 1.32 million implant-related infections are registered every year and amount to devastating consequences that expose patients to high morbidity and mortality. The implant infections are mainly associated with bacterial adhesion and biofilm formation. Bacterial adhesion and biofilm formation, in turn, are influenced by several micro-environmental factors. Some of such micro-environmental factors listed in the literature are mass transport, surface charge, surface conditioning, hydrophobicity, surface topography, and surface roughness. This chapter emphasises on the different types of implants, materials classifications, and bacteria-related infections under various sections.

1.2 CLASSIFICATIONS OF BIO-IMPLANTS

Bio-implants are prostheses devices used to regularize physiological functions. They are made up of biosynthetic materials like collagen, and tissue-engineered products like artificial skin or tissues. Most bio-engineered products like cardiac pacemakers and orthopaedic artificial implants are also covered under bio-implants because they are implanted entirely in the patient's body. Bio-implants are mostly classified at a broader level such as orthopaedic, cardiovascular, dental, ophthalmic, and neurostimulation implants; these are listed in Table 1.1.

1.3 MATERIALS USED IN IMPLANTS

The materials used in the implants should be a biocompatible, corrosion resistance, and wear resistance; they should have excellent mechanical properties and better Osseo-integration; and should not create any effect on biological system/tissue (Mahajan and Sindhu, 2018). All the available materials on earth cannot be used as biomaterials. Researchers have developed plenty of materials that can be used as biomaterials and are continually working towards developing new biocompatible materials. Some of the factors affecting implant biomaterial are *chemical factors*: these include three basic types of corrosion: general, pitting, and crevice; *surface specific factors:* the events at the bone-implant interface can be divided into the behaviour of the implant material, the host response; *electrical factors*: physiochemical methods, morphologic methods, and biochemical methods; *mechanical factor:* modulus of elasticity, tensile or compressive forces, and elongation and metallurgical aspects. The biomaterials are classified under metals, ceramics, and polymers. Some of the major materials used in the implants are listed as follows:

TABLE 1.1   Broad classifications of bio-implants (Global Trends & Forecasts till 2017)

Implant system	Different types under each implants system
Cardiovascular Implants	• Pacing devices
	• Implantable cardiac pacemakers
	• Stents and related implants
	1. Coronary stents
	2. Peripheral stents
	3. Stent-related implants
	• Structural cardiac implants
	• Heart valves
	• Tissue heart valves
	• Mechanical heart valves
	• Implantable heart monitors
Spinal Implants	• Thoracolumbar implants
	• Intervertebral spacers
	• Machined allograft spacers
	• Cervical implants
	• Implantable spinal stimulators
Orthopaedics and Trauma	• Reconstructive joint replacements
	• Knee replacement implants
	• Hip replacement implants
	• Extremities
	1. Shoulder implants
	2. Elbow replacements
	3. Ankle implants
	4. Other joint replacements fusion products

(Continued)

TABLE 1.1   (Cont.)

Implant system	Different types under each implants system
Dental Implants	• Orthobiologics
	• Trauma implants
	• Sports medicine
	• Dental screw
	• Abutment
	• Crown and bone graft
Ophthalmic Implants	• Intraocular lens
	• Glaucoma and other lenses
Neurostimulators Implants	• Cortical stimulators
	• Deep brain stimulators
	• Sacral nerve stimulators
	• Spinal cord stimulators
	• Vagus nerve stimulators

Metals: Titanium and its alloys are the metals of choice for dental and orthopaedic implants. Some of the commonly used metals are stainless steel, titanium and titanium alloys, gold, cobalt-chromium alloys, zirconium, niobium, tantalum, and so on.

Ceramics: Ceramics are non-organic, non-metallic, and non-polymeric materials manufactured by compacting and sintering at elevated temperatures. The entire implant may be made of ceramic, or this may be applied as a coat to a metallic core. Some of the ceramics biomaterials are aluminium oxide, zirconium oxide, hydroxyapatite, tricalcium phosphate, tetracalcium phosphate, calcium pyrophosphate, fluorapatite, brushite, carbon glass, pyrocarbon, and bioglass.

Polymers: Polymers are most commonly used for liners and small implants. These materials can be used alone or mixed with some other ceramics. Commonly used polymers are polyethylene, polyamide, ultra-high molecular weight polyethylene (UHMW-PE), polymethyl methacrylate, polypropylene (PP), polytetrafluoro-ethylene, silicone rubber (SR), and polyurethane (Sykaras et al., 2000).

1.4 IMPLANT FAILURE – INFECTIONS

Orthopaedic joint implants such as hip, knee, shoulder, ankle, and elbow prosthesis are used for hard tissue replacement. These are load-bearing joints subjected to a high level of mechanical stress, fatigue, and wear in the normal daily activity of humans. The artificial implants used for these applications should possess structural integrity as well as surface compatibility with the surrounding biological environment for prolonged survivability without complications and damage (Choudhury et al., 2017; Lysaght and O'Loughlin, 2000; Navarro et al., 2008; Ribeiro et al., 2012). Most of the dental implants also fail due to poor primary stability, bacterial infection, manufacturing defect, and improper selection of surgical protocol. Despite the adoption of advanced technology in man-ufacturing these new implants and in surgical and medical management procedures, the number of revisions of total knee and hip arthroplasties keeps on increasing over the period (Moriarty et al., 2016).

A significant cause for the implant failure is a fracture, prosthetic dislocation, loosening, excessive wear rate at mating surfaces and its associated debris, and pre-surgical contamination/infection (i.e., bacterial adhesion). American Joint Replacement Registry annual report (2018) collected between 2012and2017 showed that 8.2% of 47,378 hip arthro-plasties and 7.9% of 40,488 knee arthroplasties are due to the infection and inflammatory reactions. The National Joint Registry for England, Wales, Northern Ireland, and the Isle of Man surgical data of 2018 report shows that the overall 0.72% of hip replacement, 0.93% of knee replacement, and 6% of shoulder replacement results in implants-related infections. Simi-larly, it was reported in the Australian Orthopaedic Association National Joint Replacement Registry (AOA NJRR) annual report (2018) that, overall, less than 1% of revision of knee and hip arthroplasties accounts due to infection.

1.5 BACTERIAL ADHESION AND BIOFILM FORMATION

Bacterial infections on implants are usually initiated through the adhesion of bacteria to the implant surface by means of physiochem-ical interaction between the implant surface and bacteria, that is van der Waals forces, electrostatic forces, and gravitational forces (Arciola et al., 2015; Koseki et al., 2014). The reversible bacterial adhesion is followed by colonization and formation of biofilm that forms a layer of bacteria binding irreversibly to the implant surface that is later difficult to remove from the implants (Chan et al., 2017; Koseki et al., 2014). These adhered bacteria further forma colony of the bacteria that later on forms a thick layer of the dense mass of the bacteria called biofilms.

The bacterial adhesion on the implant surface depends on the

1. Properties of the implant material such as surface topography, surface roughness, surface chemistry, and surface energy;

2. Properties of the bacteria such as surface charge, surface hydrophobicity (contact angle, $\theta > 90°$)/hydrophilicity (contact angle, $\theta < 90°$); and appendages; and

3. Properties of the surrounding environment, such as the type of antibiotics and its associated flow condition, temperature, pH value, a period of exposure, chemical treatment, and bacterial concentration. Figure 1.1 shows the schematic diagram of bacterial adhesion and the effects of implant material properties.

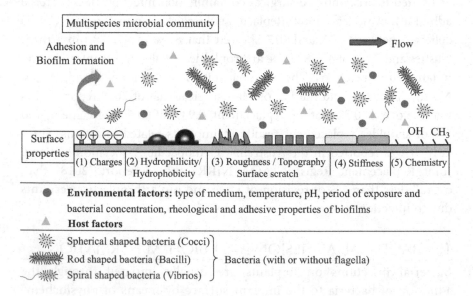

FIGURE 1.1   Schematic diagram of bacterial adhesion and the effect of implant material properties (Song et al., 2015).

1.6 EFFECT OF SURFACE ROUGHNESS ON IMPLANTS INFECTION AND WETTABILITY

In particular, surface roughness and topography are the most influencing parameters because studies show that surface roughness increases the surface area of the material, in turn, increases bacterial adhesion. Here, the focus has been made on the relation between the surface roughness and the bacterial adhesion, which is critically important from a clinical perspective when the work surfaces are finished with the finishing process. The depression, grooves, pits, scratches, and crevices in the rough surfaces influences the bacterial adhesion and acts as favourable sites for colonization and biofilm formation (An and Friedman, 2000; Barbour et al., 2007; Cox et al., 2017; Hocevar et al., 2014; Ribeiro et al., 2012; Wassmann et al., 2017; Yoda et al., 2014). These topographical features depend on the type of machining, finishing, coating, and surface treatment procedure followed. The near-net-shape of the bio-implants with the required surface finish is achieved by various finishing steps after the primary machining process. After the pre-fabrication, implants are subjected to the grinding process at the end. Final finishing of these surfaces is performed using either polishing integrated with multi-axis CNC machines or vibratory abrasive polishing, free abrasive polishing, fixed abrasive polishing, belt polishing, and so on. These processes have all been comprehensively used in the past decades (Bohinc et al., 2016; Cox et al., 2017; Kang and Fang, 2018; Turger et al., 2013) to remove the finer irregularities. It was estimated that polishing processes typically accounted for 10% to 15% of the total manufacturing cost and detailed review on different manufacturing and finishing processes used in implant manufacturing are detailed in a study conducted (Petare and Jain, 2018).

It is understood that, despite the adoption of advanced technology in manufacturing the implants and in surgical and medical management procedures, there are still a large number of implants that are subject to failure due to the surface finish achieved on the implant surfaces. The major causes for implants failure are a fracture, prosthetic dislocation, loosening, excessive wear rate at mating surfaces and its associated debris, and pre-surgical contamination/infection (i.e., bacterial adhesion). Surfaces roughness is one of the factors that influence the bacterial adhesion on implants and acts as favourable sites for colonization and biofilm formation leading to prosthetic implant infections (PIIs). A further effect of surface roughness on the wettability of surfaces is explained subsequently.

Wettability of the surface is strongly influenced by surface energy, surface roughness, and liquid surface tension. Among these, the surface roughness of any work surfaces is the most critical parameter considered in many applications and processes, where the wetting and the adhesion play a significant role. Some of the critical applications are lubrication, implants, pharmaceutical manufacturing, oil recovery, pharmacy, adhesive, coating, semiconductor, spray quenching, textiles, paper printing, polymers, dental, optics, medical field, preservation of the building, biology, cosmetics, high performance microfluidic devices, self-cleaning surfaces, and biomedical products (Guo et al., 2014; Gui et al., 2018).

Wettability is the ability of the solid surfaces to be wetted when it is in contact with any liquids by reducing liquid surface tension such that the liquid spreads over the surface and wets it. It can be described quantitatively by measuring the tangential contact angle between a liquid droplet and a solid surface-air interface. If the contact angle (θ) is less than 90^0 ($\theta < 90^0$ = hydrophilic surface), it shows that the surface has high wettability characteristics; and if it is greater than 90^0 ($\theta > 90^0$ = hydrophobic surface), it shows low wettability (Law and Zhao, 2016). Figure 1.2 shows the schematic nature of contact angle for different work surfaces (Nuraje et al., 2013).

There are many well-developed processes to modify surfaces, such as the following:

1. Non-deforming methods – polishing, machining, acid etching, and anodizing.

2. Deforming methods – sandblasting, shot peening, and surface mechanical attrition treatment.

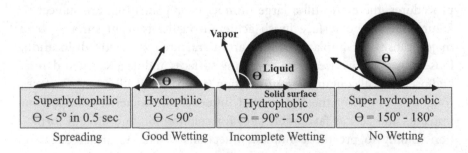

FIGURE 1.2   Nature of contact angle for different work surfaces.

The above methods are presently in use to control the roughness and wettability of metal surfaces. All these treatments control the roughness and wettability of surfaces but alter the strength of materials due to the formation of residual stresses and fine grains (Arifvianto et al., 2012).

Wenzel (1936) was the first to describe the wetting concept in a solid substrate having rough surfaces using a thermodynamic argument. The tilting plate principle was used to measure the contact angle, and he stated that the wetting properties of a solid substrate are directly proportional to the roughness of the surface wetted. Similarly, Cassie and Baxter (1944) extended the research on the analysis of wetting of porous surfaces. It is reported that when a liquid wets the porous surfaces, air pockets are formed between surface and liquid, and the interface becomes a composite surface. Based on the earlier-mentioned models, many researchers have studied the wetting of rough surfaces experimentally and analytically. Kubiak et al. (2011) established a mechanism of wettability versus surface roughness of different engineering surfaces. In the study conducted by Kubiak et al. (2011), several engineering materials such as aluminium alloy: AA7064; titanium alloy: Ti-6Al-4V; steel: AISI 8630; copper alloy, and ceramic and poly-methylmethacrylate surfaces are polished with a different process, and four different roughness values are obtained. Further, these surfaces are subjected to contact angle measurement, and conclusions are drawn based on the experimentation. It is observed that roughness has a strong influence on the wettability of engineering surfaces, and the same has been found on different materials also. Fischer et al. (2014) carried out the wettability study on the steel surfaces and reported that the directional spreading and the wettability of the surfaces are strongly influenced by anisotropic surface properties, and droplet takes place along the grooves, which leads to droplet elongation in one direction.

Li et al. (2015) experimentally identified the wetting of irregularly micro-structured surfaces. The silicon surfaces are machined with rubbing, grinding, and polishing with a 0.22 μm–3.58 μm depth, and wetting properties were investigated. This study concluded that the contact angle increases with an increase in the roughness factor and the aspect ratio on the irregular micro-structured surfaces. Also, anisotropic wetting properties were observed on regularly micro-grooved surfaces but not on the irregularly micro-grooved surface. Liang et al. (2015) conducted the wettability experiments on the aluminium alloy – AA6061 and nylon GF30 polished with ten different grit sandpapers to identify anisotropic wetting characteristics versus roughness on machined surfaces. It was concluded that roughness has a significant influence on anisotropic wetting on machined surfaces. As the roughness decreases, anisotropic wetting reduces. In the direction perpendicular to lay energy, a barrier exists, which is the main reason for anisotropic wetting. Gui et al. (2018) carried out the wettability experiments on titanium-coated textured surfaces. This study helps in understanding the anisotropic wettability of the surfaces and its applications in different areas, including orthopaedic titanium implants.

From the above literature, it is concluded that the wetting of the surface finds wide applications in many areas. Surface roughness has a great influence on the wettability of the surfaces; especially roughness range and topographical features (i.e., lay direction, width, depth, etc.) decide the anisotropic wetting behaviour of the surfaces. Further study in this direction on work surfaces finished with a different technique and how these surfaces affect the wetting is essential to understand.

1.7 SUMMARY

This chapter gives a glimpse of different types of implants and materials used in the manufacturing of implants, and also implants' failures are explained in detail. Bacterial infections on the implants are explained along with the biofilm formation on the implants. Further, broad literature on the effect of surface roughness on the bacterial infection and wettability of the implants materials surfaces are elaborated in this chapter.

CHAPTER 2

Surface Modification Techniques

2.1 INTRODUCTION

Surface properties of implants hold great importance in the response of biomaterials to the host. In this chapter, objectives of the surface modification techniques and different classifications of the surface modification methods to alter the surface properties of the implants are described. Protein adsorption, cell adhesion ability, biocompatibility, biomimetics, biodegradation, and hydrophobicity/hydrophilicity are some of the required surface properties of implants used in the biomedical applications. Surface modification is the physical alteration of the surface topography or attachment of various ligands or molecules to bring forth distinct physical, chemical, or biological properties. Surface modification can be achieved either by modifying surface atoms/molecules via physical/chemical route or by depositing a layer over the surface (Alam et al., 2015).

2.2 OBJECTIVES OF SURFACE MODIFICATION

The implant surface is mainly interacting with the various components of the biological system, such as tissue, body fluid, bone, and blood. The surface properties and topography have more influence, and to achieve these properties, researchers are continually trying to develop the various surface modification techniques. The main objective of the surface modification of biomaterials is to improve interfacial properties such as wettability, adsorption of proteins, and ligands and to improve roughness that will enhance the biomaterials' effectiveness (Alam et al., 2015).

Figure 2.1 shows the various objectives of the surface modifications used to alter the surface properties of the implants, and these include to alter the surface roughness and topography to improve the tissue growth on the implant surface; to alter the surface topography, which, in turn, reduces the bacterial adhesion, which reduces the cases of the implant failure due to biofilm formation; to improve the lubrication properties in the implant used in the joint implants; to enhance the blood compatibility such as surface should not be adherent and thrombogenic, and to increase or decrease the hydrophobicity and hydrophilicity property of the surface that decide the adhesion properties.

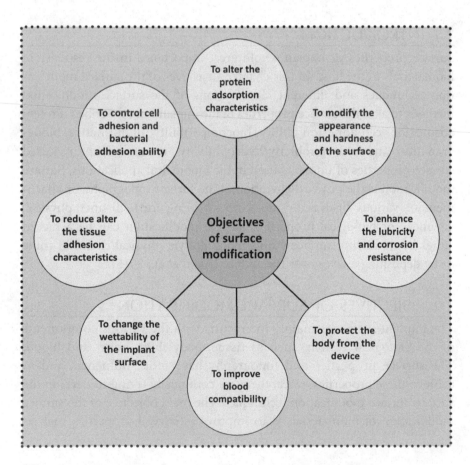

FIGURE 2.1   Objectives of the surface modification of bio-implants.

2.3 SURFACE MODIFICATION TECHNIQUES

As discussed in the introduction section, various surface modification techniques are currently in use to enhance the surface properties of the implants (Figure 2.2), and these techniques are broadly classified under three classes:

Mechanical: Mechanical surface modification methods are the most commonly used surface modification techniques, which have been used to alter the surface properties by changing the surface roughness and morphology to enhance the surface properties. Common mechanical surface modification methods are micro-machining, grinding, polishing, and blasting, which involves the physical treatment, shaping, or removal of the materials from the surface. The typical objective of mechanical modification is to obtain specific surface topographies and roughness, remove surface contamination, and improve adhesion in subsequent bonding steps.

Physicochemical: Physicochemical characteristics such as surface free energy, surface charge, chemical composition, and surface wettability are fundamental parameters that influence osteoblast attachment. Physicochemical properties can be altered primarily through manufacturing processes. Some of the chemical methods include chemical treatment,

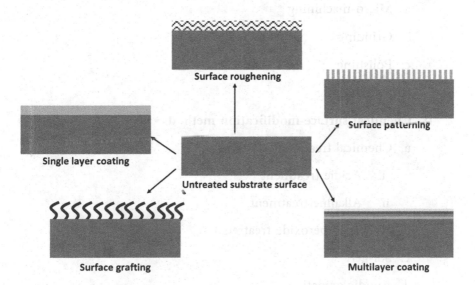

FIGURE 2.2   Basic classification of surface modification techniques.

electrochemical treatment (anodic oxidation), sol-gel, chemical vapour deposition (CVD), and biochemical modification. During the chemical treatment, electrochemical treatment, and biochemical modification, chemical, electrochemical, or biochemical reactions occur, respectively, at the interface between the biomaterial and a solution.

Physical: During some surface modification processes, such as thermal spraying and physical vapour deposition, chemical reactions do not occur. In this case, the formation of the surface modified layer, films, or coatings on biomaterials is mainly attributed to the thermal, kinetic, and electrical energy. This method includes thermal spray, physical vapour deposition (PVD), ion implantation and deposition, and glow discharge plasma treatment.

2.4 TYPES OF SURFACE MODIFICATION TECHNIQUES

Surface modification techniques used to alter the bio-implant surfaces to improve their surface properties are broadly classified under different categories. Some of the commonly used surface modification techniques (adopted from Subramani et al., 2018) are listed and explained as follows under various categories:

I. Mechanical surface modification methods

    a. Micro-machining

    b. Grinding

    c. Polishing

    d. Sandblasting

II. Chemical surface modification methods

    a. Chemical treatment

        i. Acidic treatment

        ii. Alkaline treatment

    b. Hydrogen peroxide treatment

    c. Sol-gel

    d. Anodic oxidation

e. Chemical vapour deposition (CVD)

f. Biochemical methods

   i. Osteoinductive biomolecular cues

   ii. Micro-scale and nanoscale coating of hydroxyapatite/calcium phosphate/alumina coatings with bioactive molecular cues, osteoinductive growth factors, and anti-bacterial drugs

   iii. Organic nanoscale self-assembled monolayers (SAMs)

   iv. Bioactive, biodegradable hydrogels

   v. Antibacterial agents or antibacterial drug delivery directly from titanium surface

III. **Physical surface modification methods**

a. Thermal spray

   i. Flame spray

   ii. Plasma spray

   iii. High velocity oxy-fuel (HVOF)

   iv. Detonation-gun spray (DGUN)

b. Physical vapour deposition (PVD)

   i. Evaporation

   ii. Ion plating

   iii. Sputtering

c. Ion implantation and deposition

   i. Bean-line ion implantation

   ii. Plasma immersion ion implantation (PIII)

d. Glow discharge plasma treatment

Micro-machining: Micro-machining is one of the mechanical surface modification methods in which the material is removed from the surface in the form of chips, varying from 100 µm to 999 µm. Commonly used micro-machining techniques to modify the surfaces are micro-milling, micro-electro discharge machining, micro-electro chemical machining, ultrasonic machining, micro-electro chemical discharge machining, and so on. This machining is usually performed after the bulk machining process or can be used directly to get the final shape of the component with required surface topography. Type of bulk materials, surface/material properties, machining type, machining parameters, and duration are some of the influencing parameters.

Grinding: It is a mechanical surface modification process, in which a grinding wheel acts as a mixer of abrasive particles (silicon carbide, alumina oxide, boron carbide, and diamond) and binder materials. Abrasive particles help in removing the materials from the surface and binding materials help in holding the abrasive particle together, which forms the solid-type tool of circular wheel typically used in grinding machines. The tool can be in any form based on the surface to be modified. Grinding can be used to finish the implants after machining as well as to alter the surface topography of the implant surfaces based on the application of the implants to be implanted in the biological system. In this process, grinding wheel type, grinding machine parameters, material composition, initial surface roughness, and topography have an effect on the final surface roughness to be obtained.

Polishing: Polishing is a type of mechanical finishing process used to remove the small burrs or chips left over the surface after machining or rough grinding. Typically, the polishing process is to be carried out at the end of the manufacturing process to give the final finish to the implant surface. In this process, grit paper (abrasive particles with different sizes are bonded to the paper) fixed on the rotating disc and work surface to be finished placed over this rotating polishing paper. Type of grit paper/polishing paper, initial surface roughness, type of abrasive, polishing duration, and bulk materials properties are factors to be considered during the polishing.

Sandblasting: Sandblasting, also called grid blasting, is one of the oldest and most commonly used surface modification techniques, due to the simplicity in construction and economical process compare to another process. It is generally used to improve the commercial implant osseointegration with surrounding tissue, by increasing the surface area of the implant with the help of descaling and surface roughening. Small grits in chosen shape and size are forced across implant surfaces by compressed air that creates a crater. Surface roughness is dependent on the bulk material properties, the particle size, particle shape, particle speed density, time duration of blasting, the pressure of the compressed air, and distance between the source of the particles typically nozzle exit and implant surface to be modified. Commonly used blasting media are alumina (Al_2O_3) or silica (SiO_2); surfaces blasted with 25 μm particles were rougher than the machined surface and smoother than 75 μm, and 250 μm blasted surfaces; and typical Sa values range from 0.5 to 2.0 μm. The main objectives of the sandblasting are cleaning the implant surface and increasing its bioactivity; roughening surfaces to increase effective/ functional surface area; accelerating osteoblasts adhesion and proliferation; producing beneficial surface compressive residual stress; exhibiting higher surface energy, higher surface chemical, and physical activities; and enhancing fatigue strength and fatigue life due to compressive residual stress (Subramani et al., 2018).

Sputtering: Sputtering is one of the physical vapour deposition technologies used in orthopaedic and dental implants. In this procedure, atoms or molecules of some materials are ejected in a vacuum chamber, becoming precursors for coating due to the bombardment with high-energy ions. The deposition of films is dependent on various sputtering parameters, such as the implant material properties, sputtering power and time, gas flow, and working pressure. Antibacterial agents can be effectively incorporated into implant materials by the magnetron sputtering process. The desired antibacterial ability can be preserved under optimum processing parameters (Rasouli et al., 2018; Subramani et al., 2018).

Plasma spraying: The plasma-spraying physical surface modification technique involves the projection of precursor materials into the hot plasma jet generated by a plasma torch under vacuum pressure, reduced pressure, or atmospheric pressure. Argon and oxygen are the common gases used for these applications. Upon impingement of precursor materials (powders particles) onto the implant surface, an adherent coating is formed by melting and sintering. The main advantage of plasma-spraying is the possibility of coating various nanostructured films, for example, Au, Ti, and Ag, on a wide range of materials such as ceramics, metals, or polymers at a thickness <100 nm. For example, the stream of the HA powder is blown through a very high temperature flame that partially melts and ionizes the powder, which emerges from the flame, hitting the metallic surface, which has to be coated. This method uses carrier gas, which ionizes the forming plasma and superheats the particles of HA, which undergo partial melting and are propelled towards the surface that has to be coated, producing around 50 μm thick coatings (Garg et al., 2012; Rasouli et al., 2018).

Acid etching: A variety of chemical treatments such as solvent cleaning, wet chemical etching, and passivation treatments have been employed for modifying the implant surfaces. Acid etching can remove grains and grain boundaries of the implant surface. Immersing it in strong acids (e.g., nitric acid, hydrochloric acid, hydrofluoric acid, sulfuric acid, and their mixtures) for a given period of time creates a micro-roughness of 0.5–3 μm. The surface is pitted by the removal of grains and grain boundaries of the implant surface. It also cleans the implant surface/ removes deposits. The selective removal of material and the resulting roughness are dependent on the bulk material, certain phases, the surface microstructure, impurities on the surface, the acid, and the soaking time. The etched acid implants give greater resistance in reverse torque removal and better osseointegration compared to the machined surface implants. The degree of etching is dependent on the acid concentration, temperature, and treatment time (typically from 1 to 60 min) (Rasouli et al., 2018; Subramani et al., 2018).

Alkali treatment: Alkali treatment (e.g., NaOH treatment) is a popular chemical surface treatment method. Titanium nanostructures with a sodium titanate gel layer outward from the surface have been seen after NaOH treatment. The formation of the gel-like layer over the implant surface allows for HA deposition. H_2O produces a titania gel layer. This behaviour has also been seen with other metals such as zirconium and aluminium. Alkali treatment results in the growth of a nanostructured and bioactive sodium titanate layer on implant surfaces (Rasouli et al., 2018).

Anodic oxidization: Produces roughness, porosity, and chemical composition for improved biocompatibility. The anodic oxide can have interconnected pores (0.5–2 μm in diameter) and intermediate roughness (0.60–1.00 μm). Also, the anodic oxide can be a flat layer or tubular and can have an amorphous or anatase phase. The anodization technique is commonly applied to fabricate nanostructures with diameters <100 nm on titanium implant surfaces. Anodization or "anodic oxidization" is an electrochemical deposition process carried out in an electrolyte. The deposition process is tailored by varying different process parameters to control the structural and chemical properties of the surface, including electrolyte composition, current, anode potential, and temperature (Rasouli et al., 2018).

Wet chemical deposition sol-gel: It deposits thin coatings with homogenous chemical composition onto substrates with large dimensions and complex design. In this technique, the coating is fired at 800–900°C to melt the carrier glass to achieve its bonding to the metallic substrate. The precursor of the final product is placed in the solution, and the metal implant, which has to be coated, is dipped into the solution and is withdrawn at a prescribed rate. It is then heated to create a more dense coating. Wet chemical deposition methods are alternatives to physical deposition methods, which preserve biomolecule activity. One of the most important advantages of wet chemical deposition is that drugs can be incorporated into the coatings. It has several benefits, including the simplicity of the experimental setup, mild chemical conditions of preparation, and the possibility to coat a complex 3Dgeometry onto the implants. Biomimetic modification is one of the coating techniques used to obtain successful osseointegration. The classical biomimetic coating, for example, Ca–P coating, typically requires an immersion period ranging from 14 to 28 days with replenishment of simulated body fluid (SBF) (Garg et al., 2012; Rasouli et al., 2018; Subramani et al., 2018).

Chemical vapour deposition (CVD): CVD has been used to deposit diamond nanoparticles on Ti dental implants to provide ultrahigh hardness, enhanced toughness, and good adhesion. CVD differs from PVD in the processes employed; it only uses chemical bonding to deposit the layer while PVD uses physical forces. CVD utilizes a mixed source material while PVD utilizes a pure source material. For CVD, the precursor eventually decomposes and leaves the desired layer of the source material in the substrate. HA is a bioactive ceramic with a crystal structure similar to that of a native bone and teeth minerals (Rasouli et al., 2018; Subramani et al., 2018).

2.5 SUMMARY

The surface of the implant is the main deciding factor when the biocompatibility of the implants is discussed. The implant surface should have certain properties to be compatible with the biological system and to enhance the implant function. To achieve these properties, some surface modification techniques have been developed based on mechanical, chemical, and physical modification process. These classifications are discussed in detail with examples. Before the classifications, the main objectives of the surface modification techniques are discussed. In the next chapter, more focus has been made on the mechanical-type surface modification process, which comes under the polishing process.

CHAPTER 3

Abrasive Flow Finishing

Introduction and Literature Survey

3.1 INTRODUCTION

Revolution in manufacturing has been taking place since World War II, which allows the manufacturers to meet the demands imposed by increasing the sophisticated designs but difficult to machine the materials. The typical manufacturing process – which converts the raw material into the finished product – follows the following two stages: the material conversion process and the finishing process (Hashimoto et al., 2016). The material conversion process starts with forging and forming of the raw materials and then machining of these materials to get the desired geometry. The next stage is heat treatment to alter the microstructure and mechanical properties of the components. These components are further processed through a series of finishing processes, such as grinding, and superfinishing processes, to achieve the required dimensional accuracy, tolerance, and surface topography. Surface quality and its improvement are key technology drivers for finishing processes. Over the years, surface finish requirements on the components have increased significantly. The result is the introduction of advanced manufacturing processes used for material removal, forming, and joining of various engineering materials, known today as unconventional or non-traditional manufacturing processes. These processes have the capability of creating features on work surfaces that cannot be met with conventional techniques of manufacturing. In any manufacturing process, surface formation and material removal are influenced by the geometry and the motion of the tool (Heisel and Avroutine, 2001).

Also, the demand for components with complex internal and external cavities and miniaturized products is increasing continuously in the industry like aerospace, automotive, biomedical, radar, tools, moulds, dies, defense, and turbines applications (Petare and Jain, 2018). These products require a high-quality surface finish and are mostly manufactured by unconventional manufacturing processes. Finishing these miniaturized products is a value-adding activity by which the surface texture gets modified to perform better when it is put into operation. In particular, the surface roughness has an influence on friction, wear, corrosion, and formation of cavitations, where the surfaces are subjected to sliding motion, corrosive environment, and fluid flow. Conventional finishing processes like grinding, lapping, and honing processes are most commonly used in various industries for finishing of surfaces after the machining process. Due to the development of complex geometrical shapes of engineering components, the available conventional processes are incapable of producing the required surface finish because of process limitations to access the complex features. The newer development in advanced finishing processes like abrasive flow machining (AFM), magneto rheological AFM, electro-chemical aided AFM, rotational AFM, centrifugal force-assisted AFM, etc. are used to finish the complex internal and external surfaces. These components are made of difficult-to-machine materials like non-ferrous alloys, superalloys, ceramics, refractory materials, carbides, semiconductors, quartz, composites, etc. finished efficiently and economically. The objective of the AFM process is to produce the required finish on the machined components, which are having complex intricate shapes and profiles without any difficulty. This chapter details the introduction to different finishing processes.

3.1.1 Super Finishing Processes

The final shape of the component is achieved by various finishing steps after the primary machining processes. These finishing processes are classified into the following two types: motion copying processes (grinding) and pressure copying processes (abrasive fine finishing). Figure 3.1 shows the classification of abrasive finishing processes. In the motion-copying processes, the depth of material to be removed is given as an input to the machine. The best example for this type of process is grinding in which fixed depth of material removed for each rotation of

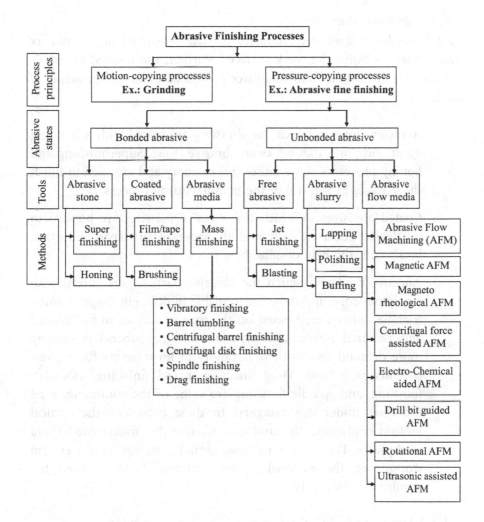

FIGURE 3.1 Classifications of abrasive finishing processes (Hashimoto et al., 2016).

the bonded abrasive wheel. But this process cannot be used to finish complex internal features and, moreover, it can induce thermal and mechanical stresses on the work surfaces (Petare and Jain, 2018).

In the pressure copying processes, no fixed depth of cut is given, and instead of this, the material is removed due to the pressure of the tool against the work surfaces (Hashimoto et al., 2016). Further, pressure copying processes are divided into two states – (a) Bonded and (b) Unbonded. These are explained as follows.

3.1.1.1 Bonded Abrasives

In a bonded abrasives state, abrasive particles are fixed on the matrix that is used to finish the work surfaces. Further, the bonded abrasives are subdivided into the following three groups based on the principle of work:

a. Abrasive stone in which the abrasives are mixed with a bonding agent and pores that act as an abrasive stone. Superfinishing and honing processes come under this category and are used to finish the surfaces at the end of the manufacturing process.

b. Coated abrasives in which the abrasive particles are bonded to the flexible substrate such as cloth, paper, plastic tapes, and nylon brush with the help of adhesives.

c. Abrasive media in which the abrasive particles are mixed with a polymer core, and these are available in different shapes (cones, triangle, sphere, etc.) based on the type of surface to be finished and required finish. This type of finishing process is gaining more demand due to their flexibility in finishing multiple components at a time. Drag finishing, barrel finishing, vibratory finishing, and spindle finishing are some of the commonly used processes under this category. In these processes, the cyclical motion is given to the container filled with abrasive media and workpieces. Due to the rubbing motion between abrasives and workpieces, the material is removed and hence formed the required surface finish.

3.1.1.2 Unbonded Abrasives

In an unbonded abrasives state, abrasive particles freely participate in the finishing process, and the viscosity of these abrasive carriers varies with the type of carrier used, such as gas, fluid, and solids (flexible). Further, unbonded abrasives are subdivided into three groups.

a. Free abrasive in which abrasives are forced towards the surface to be finished with the help of jet of air. Examples: abrasive jet finishing and abrasive blasting.

b. The abrasive slurry in which abrasives are mixed with the fluid, which is used to finish the surfaces. Example: lapping and honing in which slurry is introduced between the workpiece and tool. The pressure is applied on the workpieces that help in removing the surface irregularities. Buffing falls under this group, in which abrasive slurry is applied on the buff that rotates at higher speed, and workpieces are fed against to this wheel. These processes are very slow and restricted to simple work surfaces.

c. Abrasive flow media, in which the abrasive particles are mixed with flexible polymer media and extruded over the surfaces to be finished. This process is used to finish complex internal and external surfaces which are difficult to access with other finishing processes (Hashimoto et al., 2016). Figure 3.2 shows the list of different methods used to manufacture the components and surface roughness ranges achieved from these processes.

3.1.2 Surface Characteristics

The solid surfaces are boundaries between the component and its working environment (Smith, 2002). The structure and properties of the finished surfaces depend on the nature of the solid material, methods used to prepare the surfaces and interaction. Properties of the solid surfaces are getting more important in engineering because the surface properties play a major role in optical, thermal, electrical, tribological, and painting applications. Irregularities and deviations present over the surfaces cannot be avoided irrespective of the method used to produce the work surfaces. In general, no machining process exists that can produce molecularly flat surfaces. Even the smoothest surface contains irregularities in micro- or nano-scale, and both macro- and micro-/nano-surface topography are important in technological applications (Bhushan, 2013). Surface topography provides the quantitative dimensions of the features present on the surface in terms of roughness, features size, and step height. These dimensions can be measured using contact-type stylus instruments, an optical profilometer, the confocal microscopy, the interferometry, the atomic force microscopy, etc. Similarly, the surface morphology provides a qualitative evaluation of the three-dimensional shape of the surfaces evaluated by optical microscopes and scanning electron microscope. Some of the common terminologies associated with surface topography are explained in the following text.

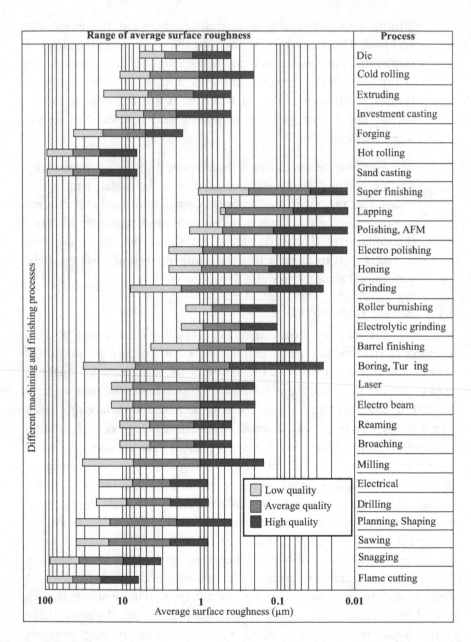

FIGURE 3.2   Average surface roughness obtained from machining and finishing processes (PC: https://www.engineeringtoolbox.com/surface-roughness-d_1368.html).

3.1.2.1 Surface Topography Terminologies

Figure 3.3 shows the schematic of the surface texture of the surface. Some of the important terminologies used when describing the surfaces are:

- *Waviness:* Waviness is the surface irregularities of longer wavelengths. It may result from factors such as vibration, chatter, machine or workpiece deflections, warping strains, or heat treatment (Bhushan, 2013; Menezes et al., 2013).

- *Roughness:* Roughness is the surface irregularities of shorter wavelengths. It is characterized by surface asperities peaks (maximum height on a portion of a profile) and valleys (maximum depth on a portion of a profile) of varying amplitude and spacing. It is a result of the manufacturing process used to convert the bulk material into the required shape. Some of the examples include the tool mark left on the surface as a result of turning and the impression left by grinding or polishing (Whitehouse, 2011).

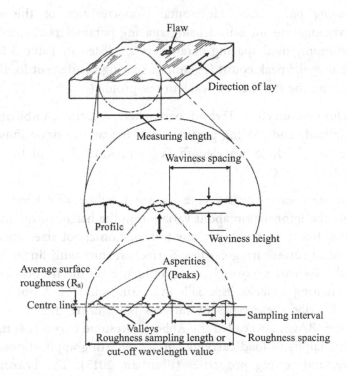

FIGURE 3.3   Schematic of the surface texture (Bhushan, 2013).

- *Flaws:* Flaws are unexpected and unintentional interruptions on the surface structure caused by material structural dislocations and other material defects.

- *Lay:* Lay is the principal direction of the predominant surface pattern produced by the type of surface modification process.

The following section includes the classifications of surface roughness parameters.

3.1.2.2 Classifications of Surface Roughness Parameters

- *Amplitude Parameters:* Vertical characteristics of the surface deviations are measured using amplitude parameters. Some of the commonly used amplitude parameters are listed in Table 3.1. They are namely: average roughness (R_a), root-mean-square (RMS) roughness (R_q), the maximum height of peaks (R_p), maximum depth of valleys (R_v), the maximum height of the profile (R_t), and ten-point height (R_z).

- *Spacing parameters:* Horizontal characteristics of the surface deviations are measured using spacing parameters. Some of the commonly used spacing parameters are listed in Table 3.1. They are namely: peak count (P_c), mean spacing of adjacent local peaks (S), and the number of peaks in the profile (m).

- *Hybrid parameters:* Hybrid parameters are the combination of amplitude and spacing parameters of the surface deviations. They are namely: RMS wavelength (λ_q) and RMS slope of the profile (Δ_q).

- *Bearing area curve:* Average surface roughness (R_a) value gives only the information about variation in the height of the irregularities, but it does not give any information about size, shape, and slope of surface irregularities. Surfaces texture with linear grooves and crisscross grooves have an influence on the friction between the mating surfaces, especially in bearing surfaces. A possible way to describe these types of surfaces is by using the bearing area curve (BAC), also called the Abbott–Firestone curve. It is used for the analysis of load carrying surfaces, bearing applications, adhesive, and sealing properties (Bhushan, 2013). The bearing area curve is the ratio of the air to metal starting from the highest peak

TABLE 3.1 Some of the important surface roughness parameters (Gadelmawla et al., 2002; Smith, 2002; Whitehouse, 2011)

	Roughness parameter	Description	Applications
Amplitude parameters	R_a: Average roughness	The average absolute deviation of the roughness irregularities from the mean line over one sampling or assessment length.	The commonly used parameter in most of the applications
	R_q: Root-mean-square roughness	The standard deviation of the surface height distribution, defined on the sampling length.	These are mainly in common use in the USA
	R_p: Maximum height of peaks	Maximum height of the profile above the mean line within the assessment length.	Used in the evaluation of frictional force and electrical contact resistance
	R_v: Maximum depth of valleys	Maximum depth of the profile below the mean line within the assessment length.	Evaluation of surface strength and corrosion resistance
	R_t: Maximum height of the profile	The vertical distance between the highest peak and the lowest valley along the assessment length of the profile.	–
	R_z: Ten-point height	The difference in height between the average of the five highest peaks and the five lowest valleys along the assessment length of the profile.	Evaluation of gloss and luster, surface strength, surface treatability, frictional force, electrical resistance
Spacing parameters	P_c: Peak count	The number of local peaks, which is projected through a selectable band located above and below the mean line by the same distance.	Forming, painting, coating surfaces
	S: Mean spacing of adjacent local peaks	The average spacing of adjacent local peaks of the profile measured along the assessment length.	Evaluation of gloss and luster, high-grade feel, adhesion performance, and surface treat
	m: Number of peaks in the profile	The number of peaks of the profile per unit length.	–

(Continued)

TABLE 3.1   (Cont.)

	Roughness parameter	Description	Applications
Hybrid parameters	λ_q: RMS wavelength	The root means of the measure of the spacing between local peaks and valleys, taking into consideration their relative amplitudes and individual spatial frequencies.	-
	Δ_q: RMS slope of the profile	The root-mean-square of the mean slope of the profile.	Evaluation of gloss and luster, surface treatability, frictional force, and corrosion resistance

and dropping down to the lowest valley on the surface profile and is generally referred to as a bearing ratio (Whitehouse, 2011). Table 3.2 lists the bearing area curve or Abbott material ratio curve and description of parameters with the profile. So the BAC is used to show the quality of surface texture and also for a measured surface; it gives access to the size and proportions of the peaks and valleys present.

3.1.3 Abrasive Flow Machining/Finishing

Abrasive flow machining (AFM) is a non-traditional machining process in which the viscoelastic polymer-based flexible media mixed with abrasives is extruded through the complex profiles to achieve the better finish and to remove the burrs left out inside profiles due to previous machining processes.

This process has the ability to finish any components having complex internal or external surfaces made up of various engineering materials like superalloys, ceramics, refractory materials, carbide, and non-ferrous alloys. This process is used to polish, radius, deburr, remove cast layers, and polish the work surfaces (Rhoades, 1991). Extrude Hone Corporation, USA, in 1960, introduced a commercial AFM machine to finish the aerospace components effectively. The material removal mechanism is explained in the following text.

3.1.3.1 Material Removal Mechanism

Figure 3.4 shows the mechanism of material removal in the abrasive flow finishing process. The material removal mechanism is almost the

TABLE 3.2   Bearing area curve or Abbott material ratio curve and description of parameters with the profile (Graf, 2015; Menezes et al., 2013)

Roughness profile	Bearing area curve

Roughness parameter	Description
R_{pk}: Reduced peak height	The top portion of the surface which will be worn away in the running-in period.
R_k: Core roughness depth	This is the working part of the surface. After the initial running-in period, it will carry the load and influence life and performance.
R_{vk}: Reduced valley depth	The lowest part of the surface, which has the function of retaining the lubricant.
M_{r1}: Peak material component	The bearing ratio at which R_{pk} and R_k meet. This is the upper limit of the core roughness profile and derived from the bearing ratio plot.
M_{r2}: Valley material component	The bearing ratio at which R_{vk} and R_k meet. This is the lower limit of the core roughness profile and is derived from the bearing ratio plot.
A1 and A2	Area of peaks and valley

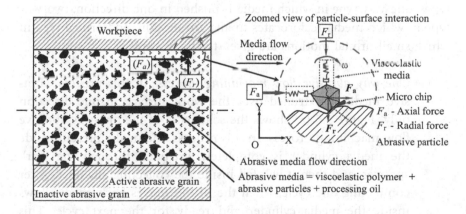

FIGURE 3.4   Mechanism of material removal (Fu et al., 2017; Gorana et al., 2004).

same as the grinding process, but the type of tool used in grinding and abrasive flow machining is different. In AFM, a flexible abrasive medium is used, which is a mixer of the viscoelastic polymer media, the plasticizer, and the abrasive particles, and this is extruded over the work surfaces to be finished. The extrusion pressure of the media produces the axial and radial forces acting on the active abrasive particle. The F_a and F_r are the axial and radial forces of the active abrasive particles acting on the surface. These forces help in removing the irregularities present over the work surface, and these can be determined with the help of the viscoelastic properties of the media. The elastic component of media assists to exert a radial force (F_r) on the abrasive particles, which, in turn, create the indentation on the work surface. Viscous component of media assists to exert an axial force (F_a) and helps to move the indented abrasive particle along the direction of applied pressure (Fu et al., 2017; Gorana et al., 2004). Both viscous (F_a) and elastic components (F_r) help in removing the materials in the form of micro- and nano-chips.

The material removal mechanism includes three deformation modes – elastic deformation, plastic deformation, and micro cutting of the material, which are influenced by cutting force and depth of indentation. The highest material removal takes place in the area with the greatest media flow restrictions (Williams, 1993). Different types of AFM based on the media flow over the work surface are detailed as follows.

3.1.3.2 AFM Types Based on the Direction of Media Flow

Based on the motion of the abrasive media, AFM is classified into three types: one-way type in which media is pushed in one direction, two-way type in which media reciprocates to and fro motion, and orbital type in which small orbital motions are given to the workpiece.

(1) *One-way abrasive flow machining process*: In this process, abrasive media is extruded over the work surface in one direction only. Figure 3.5 (a) shows the schematic of the one-way abrasive flow machining process. In this process, the media present inside the media cylinder is completely extruded through the work surfaces, and then it will fall inside the collecting chamber. After completing one cycle, when the piston moves back, media flows inside the media cylinder and ready for the next cycle. This process is mainly used to finish complex profiles such as rotors,

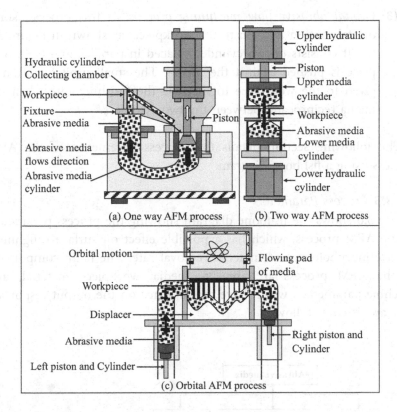

FIGURE 3.5   Types of abrasive flow machining processes (Petare and Jain, 2018).

turbine blades, components having multiple radial holes, mani-fold blocks, nozzles, injectors, etc.

(2) *Two-way abrasive flow machining process*: In this process, abra-sive media reciprocates to and fro motion over the work surface to be machined. Figure 3.5 (b) shows the schematic of the two-way abrasive flow machining process. The working is almost similar to the one-way process, but this technique consists of two media cylinders, which are placed exactly opposite to each other. In between these cylinders, the workpiece is placed in the work fixture, and the media reciprocates inside this workpiece. One complete forward and reverse movement of the piston is considered as one cycle in this process. This process is mainly used to finish the gears, the nozzles, the micro channels, the micro holes, the freeform surfaces, etc.

(3) *Orbital abrasive flow machining process*: In this process, small orbital motion is given to the workpiece, as shown in Figure 3.5 (c). It consists of two cylinders placed in parallel, and the workpiece is pushed against the media. The small orbital motion is given to the workpiece to increase the finishing rate when the media reciprocates between the two cylinders.

The following section details the process parameters of the AFM process using a fishbone diagram.

3.1.3.3 Process Parameters

Figure 3.6 shows the fishbone diagram showing the process parameters of the AFM process, which has a possible effect on surface roughness, surface morphology, and material removal rate. The main components of the AFM process are abrasive media, workpiece material, and machine parameters, which have an influence on the output responses; they are listed as follows:

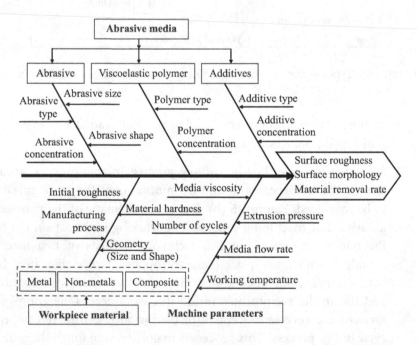

FIGURE 3.6   Fishbone diagram showing the process parameters of AFM.

(a) *Abrasive media*: Abrasive media is an important tool in the AFM process on which the whole finishing process is dependent. Better surface finish can be achieved by selecting the proper media constituents and their percentage of the composition. The media used in this process should pose better flowability, self-deformability, abrading ability, chemically inactive, and non-corrosive properties to finish the different work surfaces. Media parameters are grouped under three categories, such as abrasives, viscoelastic polymer, and additives. Under abrasive particles – abrasive type (silicon carbide, alumina oxide, boron carbide, diamond, etc.) concentration, size (60 mesh size to 2,000 mesh size), and shape are the influencing parameters. Viscoelastic polymer type, namely,polyborosiloxane, silicone rubber, natural rubber, silly putty, etc., and its concentration also have an influence on the finishing process. The additives or plasticizers, like hydraulic oil, silicone oil, etc., type and its concentration are the other parameters that affect the media viscosity and which decides the media flowability. The rheological properties of media, after blending all these individual constituents, play a major role, and they determine the performance of the media during the operation.

(b) *Workpiece material*: Type of workpiece, geometry, initial surface roughness, and type of primary manufacturing process are important parameters related to workpiece material, which have an influence on the finishing process.

(c) *Machine*: Extrusion pressure, media flow rate, and a number of cycles are the machine parameters that have a direct influence on finishing efficiency.

The next section details the literature on the AFM process and the effect of process parameters on the finishing of different engineering materials and their surfaces.

3.2 LITERATURE SURVEY

An extensive literature survey has been carried out on different aspects of AFM process, and the observations are segregated under the following six sections:

1. Experimental setups and experimentation

2. Development of abrasive media

3. Modelling and optimization

4. Application areas

5. Effect of surface roughness on wettability

6. Effect of surface roughness on implants infection

3.2.1 Experimental Setups and Experimentation

The abrasive flow machining process was first introduced by Extrude Hone Corporation, USA, in 1960 to finish the complex internal features of the components used in aerospace applications. Since then, researchers are striving hard to enhance the performance of this process and also exploring the different areas in which this process can be used to finish the complex surfaces. To enhance the performance of the traditional AFM process, many researchers have developed the hybrid machining processes in which the various machining processes are combined with the AFM process to achieve the higher material removal rate (MRR) and the required surface finish in lesser time.

Some of the recent developments in hybrid machining processes are magneto abrasive flow machining (MAFM), magneto rheological abrasive flow finishing (MRAFF), centrifugal force-assisted abrasive flow machining (CFAAFM), electro-chemical-aided abrasive flow machining (ECAFM), drill bit guided-abrasive flow finishing (DBG-AFF), rotational-abrasive flow finishing (R-AFF), ultrasonic-assisted abrasive flow machining (UAAFM), and rotational magneto rheological abrasive flow finishing (R-MRAFF). This section provides the available literature on the development of different experimental setup and observations drawn based on the experimentation conducted by the researchers.

Loveless et al. (1994) studied the effectiveness of the abrasive flow machining process in which various work materials were pre-machined by grinding, turning, wire electrical-discharge machining, and milling. On observing the roughness pattern by scanning electron microscope (SEM) images, it was found that the surface machined by wire electrical-discharge machine is improved greatly by the AFM process compared to other processes. Further, it is reported that the type of machining process used to prepare the workpieces initially and also the type of abrasive media used to finish these surfaces significantly affects the roughness improvements.

Williams (1998) developed an online monitoring strategy using an acoustic emission sensor for determining the performance characteristics of the abrasive flow machining process. RMS voltage of acoustic emission signal found to be sensitive to the applied extrusion pressure and other process parameters, which have a direct effect on change in MRR.

The dominant process parameters that affect the MRR and surface finish were examined by Jain and Adsul (2000). The experiment was conducted on brass and aluminium. The surface texture was examined using an SEM. The study showed that the abrasive mesh size, the concentration of abrasive, the working cycles, and the abrasive media flow rate are the dominant process parameters that affect the MRR and surface finish. Further, it was observed that the initial surface roughness and hardness of the materials govern the amount of material removed from the work surface.

Singh and Shan (2002) developed the MAFM process by applying a magnetic field around the workpiece to improve the MRR and to reduce surface roughness. Analysis of variance (ANOVA) technique has been used to identify the most significant parameters – magnetic flux density, volume flow rate, number of cycles, medium flow volume, abrasive grit size, abrasive concentration, and reduction ratios. Improved surface finish and improved MRR were observed in MAFM over AFM.

Raju et al. (2003) developed an extrude honing experimental setup, in which the abrasive media flows to finish the spheroidal graphite cast iron parts. Here, silicone-based polymer media mixed with silicon carbide abrasives was extruded under different velocity with varying number of cycles and observed improvement in the surface finish as the cycles and velocity increases.

Jha and Jain (2004) explored the MRAFF process for finishing complex internal geometries. In this process, magnetorheological polishing fluid consists of carbonyl iron powder, and silicon carbide abrasives are mixed with viscoelastic base grease and mineral oil used to finish stainless steel workpieces. No improvement in the surface finish at zero magnetic field condition and significant improvement in the surface finish as the magnetic field strength increases were observed.

Gorana et al. (2004) experimentally investigated the forces acting on the work surface during the AFM process. Two-component disc dynamometer was used to measure the axial and radial force, which takes care of material removal in this process. It was concluded that the extrusion pressure, the abrasive concentration, and the grain size have a significant effect on the axial and radial forces. Dabrowski et al. (2006) developed the ECAFM process by using polymeric electrolytes for smoothing flat surfaces. Experimental investigations have been carried out for the smoothing of flat surfaces using gelated polymers and water-gels based on acrylamide as polymer electrolytes. They have reported that the application of electrochemical has an effect only when appropriate abrasive paste selected was known.

Walia et al. (2006a) tried to improve the performance of the AFM process by applying the centrifugal force on the abrasive media by introducing a rotating centrifugal force generating rod in the workpiece passage. The process is termed as centrifugal force-assisted abrasive flow machining (CFAAFM). They have concluded that the better surface finish can be achieved by CFAAFM due to centrifugal action caused by the rod on the abrasive media. Also reported, 70% to 80% of the machining time is reduced by introducing the centrifugal force generating rod while machining.

Singh et al. (2008) experimentally demonstrated the material removal mechanism in the AFM process while machining the aluminium and the brass material. By carefully observing the SEM images, they have concluded that the material removal takes place in the form of micro-chips by lip formation (plowing), embrittlement, followed by fragmentation due to the flow of abrasives. This machining process is most effective on work materials with an optimum combination of high order plastic flow and high rate of work hardening properties.

Sankar et al. (2009a) conducted the abrasive flow machining experiments on metal matrix composites – aluminium alloy reinforced with a different weight fraction of SiC particles. The extrusion pressure from 4 MPa to 8 MPa was varied, and a number of cycles up to 1,000 along with different weight percentage on the oil used in media were varied to study the parameter effect. It was observed that the extrusion pressure increases up to 6 MPa and has an influence on improvement in a change in roughness (ΔR_a), and after that, change in roughness decreases. The maximum change in roughness is achieved on Al alloy/SiC (10%) material. Sankar et al. (2009b) introduced the concept of rotating medium along its axis to achieve a higher rate of finish and material removal, and this process is called as drill-bit guided-abrasive flow finishing (DBG-AFF) process. The experiments were conducted on AISI 1040 and AISI 4340 workpiece materials. Higher finishing rate and MRR were observed in the DBG-AFF process compared to the abrasive flow finishing (AFF) process.Sankar et al. (2009c) developed a rotational-abrasive flow finishing (R-AFF) process. In this process, a workpiece is rotated at a certain speed to enhance the performance of the finishing process. The 44% improvement in ΔR_a and 81.8% of more material removal were observed in the case of developed technique compared to the conventional AFM process. This is due to the existence of the tangential force because of workpiece rotation, which assists the axial and radial force.

Das et al. (2012a) proposed a rotational magnetorheological abrasive flow finishing (R-MRAFF) process to enhance the finishing performance of the MRAFF process. In this process, a rotation and a reciprocating motion were provided to the abrasive medium by a rotating magnetic field and hydraulic unit. Smooth and mirror-like surfaces with roughness in the nanometer range were observed in both stainless steel (110 nm) and brass workpieces (50 nm). It was also reported that the cross-hatch patterns on the R-MRAFF finished surfaces from the SEM images.

Gov et al. (2013) investigated the effect of workpiece hardness while finishing with the AFM process. The workpiece of AISI D2 tool steel heat treated to different hardness numbers 31, 45, and 55 HRC was used for machining. They have concluded that a workpiece with a higher hardness number takes more number of cycles to get the required surface finish. Also, it was observed that indentation marks on the softer material are more as compared to hard material.

Wang et al. (2014) carried out experiments using the AFM process with different passageways instead of the usual circular passageway to provide multiple flow paths for the abrasive media to reduce the surface roughness in through holes. The number of helical grooves was provided in the tool placed inside the workpiece. The results indicate that the helical passageway is better in reducing the roughness improvement rates by 76% (four-helix grooves) compared to 60% by a circular passageway.

Chen et al. (2014) investigated on surface roughness improvement in drilled holes using the newly developed magneto-elastic AFM concept. They have noticed that the newly developed magneto-elastic abrasive media eased the ploughing force and avoided deep scratches on the surface of the workpiece. The results showed that the viscosity of the medium should be high initially to remove material in bulk, and later the viscosity should be reduced to increase the fluidity of the medium. Ring magnet along the workpiece helps in the reduction of roughness from 0.9 μm to 0.0541 μm. Also, an 8°C increase in temperature was observed during the machining.

Sharma et al. (2015) introduced an ultrasonic assisted abrasive flow machining (UAAFM) process in which the workpiece is subjected to ultrasonic vibration perpendicular to the media flow direction. Prefinished cylindrical specimens were finish-machined using the AFM, and UAAFM modes of machining and improved surface finish were observed in the UAAFM method. Highest %ΔR_a of 80.12% was achieved on surfaces finished with extrusion pressure of 7 bar, an abrasive mesh of 300, the processing time of 7 min, and the frequency of 15 Hz, respectively.

The review of literature available based on the experimentation reveals that the AFM process is one of the best and ideal solutions to finish complex surfaces and features. Machine parameters such as extrusion pressure, number of cycles, abrasive media, and workpiece parameters have a great influence on the finishing performance. Most of the work is carried out using the two-way process, and fixture design is also a challenging task based on the features to be finished. It is reported that the hybrid two-way process may create the criss-cross texture on a surface, and also this process is heavy and costly (Petare and Jain, 2018; Sankar et al., 2009b, 2009c). Further, in the next section, literature available on abrasive media development and their characterization are reported.

3.2.2 Abrasive Media Development and Characterization

Abrasive media is one of the key elements in the AFM process. In this process, abrasive media is the self-deformable viscoelastic non-Newtonian fluid that should pose better flowability, self-deformability, abrading ability, chemically inactive, and non-corrosive properties. The medium consists of the viscoelastic polymer and the plasticizer reinforced with suitable abrasive particles. In this, the viscoelastic polymer acts as a carrier medium, and a plasticizer is added to the base polymer to reduce the viscosity of the polymer, which also improves the flowability, flexibility, and self-deformability of media (Sankar et al., 2011). Abrasive particles are added to it, which acts as a cutting tool and removes the material from the work surface. Extrude Hone Corporation, USA, is the major manufacturer of AFM media, which is expensive, and the composition of this media is not at all available (Petare and Jain, 2018). Commonly used polymer carriers are polyborosiloxane, silicone rubber, silly putty, natural rubber, styrene butadiene rubber, butyl rubber, etc., and abrasive particles are silicon carbide (SiC), boron carbide (B_4C), and alumina oxide (Al_2O_3). Some of the researchers have tried to develop alternative AFM media apart from the commercially available media, which are discussed in this section.

Hull et al. (1992) developed polyborosiloxane mixed with diamond and SiC particles abrasive media with abrasive size varying from 0.25 μm to 1,200 μm. The abrasive concentration of 66% is used in the media, with the remaining percentage of the polymer media. The major conclusions drawn in this study are: (a) stick-slip phenomenon coupled with medium compressibility are important characteristics and have an influence on finishing, (b) due to the time-dependent behaviour of the base, media control and optimization of process are difficult, and (c) temperature rise during the machining is beneficial in rapid media flow through the small orifice. Trengove (1993) investigated the thermal and rheological properties of commercially available abrasive media based on polyborosiloxane mixed with SiC particles procured from the Extrude Hone Corporation. Three different kinds of abrasive media, high, medium, and low viscosity, are considered in this study. The media shows the pseudoplastic nature, and at shear strain rate less than 100 s^{-1}, the pseudoplasticity is low and shear viscosity may be approximated to 10 Pa.s. The pseudoplasticity nature increases with an increase in the shear rate of more than 100 s^{-1}. It also reported that the viscosity of the media increases when it is forced in the small restrictions due to an increase in shear stress and shear strain rate during this time abrasive media acts as a solid-like state.

Jain et al. (2001) developed abrasive media by mixing putty (sealant material), abrasive particles, and varnish oil (additives) with different compositions and abrasive mesh size. The effect of abrasive mesh size, concentration, and temperature of media on the viscosity of medium have been reported and concluded that these three parameters have more impact on media viscosity.

Tzeng et al. (2007a) developed a self-modulating abrasive medium whose viscosity can be adjusted during the preparation of the media to remove the recast layer from the micro-channels produced by the wire-EDM process. The media developed is a mixer of polymer, wax, silicone oil, and SiC abrasive grains. They have concluded that the AFM process improves the quality of the channels by removing burrs, straightness, recast layer, etc., with low cost and high efficiency.

Wang et al. (2007) developed two new abrasive media – pure silicone rubber P-Silicone and A-Silicone mixed with additives SiC particles in the concentration of 50%. Improved surface finish is observed when components are finished with A-silicone abrasive.

Kar et al. (2009a) developed and studied the performance of five different types of polymers-based abrasive media: (a) natural rubber, (b) ethylene propylene diene monomer, (c) butyl rubber, (d) silicone rubber, and (e) styrene butadiene rubber mixed with SiC abrasive particles. Out of five media styrene butadiene, rubber-based media shows better performance in terms of viscosity, thermal stability, and finishing of work surfaces. They have conducted a study on commercial media and found a composition of 66% abrasives, 34% carrier, and other ingredients in the commercial media using thermogravimetric analysis. Kar et al. (2009b) used two different rubbers for media development – (a) natural rubber + SiC abrasive particles and (b) butyl rubber + SiC abrasive particles with naphthenic oil as processing oil. The rheological properties of the media and the effect of these media on the finishing process were studied. Based on the experiments conducted, Butyl rubber-based abrasive media had shown good performance compared to natural rubber-based media. Sankar et al. (2011) developed styrene-butadiene-based media to finish the aluminium-based metal matrix composites. They have also investigated the effect of rheological properties of abrasive media on the finishing and the obtained results are explained with respect to the media properties.

From the literature, it is observed that abrasive media parameters such as polymer type and concentration, plasticizers type and concentration, and abrasive particle type, concentration, shape, and size are the most influencing parameters. These are blended with definite proportions to achieve better thermal, mechanical, and rheological properties, which in turn affect the finishing process. But very few papers are available on media development and their complete characterization study, and this area is still open for future research work. In the next section, some of the recent literature based on the modelling and the optimization of the AFM process are elaborated.

3.2.3 Modelling and Optimization

Modelling, simulation, and optimization are the techniques used to study the effect of process parameters on the output responses – surface finish and MRR. Some of the researchers developed the mathematical modelling and FEM simulation of the AFM process.

Apart from the FEM simulation and modelling, the different statistical models, along with the design of experiments (DOE) and optimization tools, have also been used by many researchers. Some of them are ANOVA, Taguchi, grey relational analysis (GRA), response surface methodology (RSM), central composite design (CCD), artificial neural network (ANN), genetic algorithms (GA), group method of data handling (GMDH), data dependent systems (DDS), Pareto-based optimization techniques, etc. In this section, the research findings on modelling, simulation, and optimization of process parameters are listed in the following text.

Williams and Rajurkar (1992) used a stochastic modelling and analysis technique called DDS to study the AFM-generated surface. The effect of input parameters such as media viscosity, extrusion pressure, and a number of cycles on the output responses like MRR and surface finish are discussed in the paper.

Petri et al. (1998) developed the predictive process modelling – neural networks to determine the effect of a set of process parameters on surface finish. The process parameters are characterized in five categories – workpiece parameters, media parameters, machining parameters, technical specification parameters, and process objectives parameters. This model mainly reduces the development time for new applications of the process and gives the information on the effect of input variables on output parameters.

Jain et al. (1999) developed a FEM to evaluate the stresses and forces developed during the machining process. A theoretical approach was also proposed in the paper to estimate the MRR and the surface finish during the machining process. The theoretical results were compared with the available literature on experimentation, and they were found to be in agreement with the published literature.

Jain and Jain (1999) evolved a versatile simulation model to predict the surface roughness and the MRR with reference to the abrasive size and concentration. The predicted results and the RSM results are compared to understand the relative importance of AFM parameters.

Jain and Jain (2001) proposed a theoretical model to determine the specific energy and tangential forces acting on the AFM process based on five main parameters – grain size, applied pressure, the hardness of workpiece, number of active grains, and number of cycles. By considering the heat flows to the workpiece and the medium, one-dimensional heat transfer analysis has been carried out to determine the change in temperature of the workpiece. For the prediction of the active grain density, the concept of stochastic methodology was introduced by Jain and Jain (2004), which generates and statistically evaluates the interaction between the abrasive grains and the work surface. They have concluded that the grain density increases with an increase in the mesh size and the abrasive concentration.

Ali-Tavoli et al. (2006) were the first to introduce a GMDH-type of neural networks and generic algorithms for modelling the effects of the number of cycles and abrasive concentration on both MRR and surface finish. These neural network models are then used for multi-objective Pareto-based optimization of AFM, considering two conflicting objectives, such as MRR and surface finish. A combination of GMDH-type of neural network modelling and multi-objective Pareto optimization approach is very much promising in discovering the useful and interesting design relationships.

Taguchi's parameter design strategy has been applied to investigate the effect of process parameters of the centrifugal force-assisted AFM process on MRR and surface roughness by Walia et al. (2006b). Mali and Manna (2010) used the Taguchi experimental design concept L_{18} ($6^1 \times 3^7$) mixed orthogonal array to determine the S/N ratio and to optimize the AFM process parameters. ANOVA and F-test values indicate the significant parameters affecting the AFM finishing performance.

Das et al. (2012b) carried out the computational fluid dynamics simulation and experimentation to understand the mechanism of the magnetic field-assisted nano-finishing process. Non-Newtonian model – Papanastasiou's modified Bingham plastic viscosity model is used in the paper. The axial and radial forces acting on the work surfaces are predicted using CFD simulation.

Shear thinning and viscoelastic behaviour are the major reasons for complexity in the simulation of the abrasive flow machining process, and also, a tool used in this process (or the media) is deformable. Uhlmann et al. (2013) developed a theoretical material model for the viscos-elastic abrasive medium used in the AFM process using standard Maxwell model and Generalized Maxwell model. They have assumed that material removal is being caused by shear stress dependent on the bonding of the abrasive grains. These models are validated with experimental results.

Sushil et al. (2015) designed L_{27} orthogonal array for optimizing the process parameters to machine Al/SiC metal matrix composites. They have concluded that the extrusion pressure is the most significant factor in regards to MRR and changes in average roughness value.

So, this section contained the published papers on modelling, simulation, and optimization of the AFM process, which reveals that various researchers have developed a model based on the numerical and soft computing methods. Due to the complex properties of media, it is difficult to model the process that can predict the output responses near to the experimental value. But the soft computing models based on the experimentations can be used to predict the responses within the experimental limit.

3.2.4 Applications

The researchers have made an attempt to develop the AFM technique for finishing the various components that have complex internal cavities and the external surfaces of various engineering materials, as shown in Figure 3.7. These materials are commonly used in MEMS, industrial, aerospace, automotive, and biomedical applications. Using the AFM

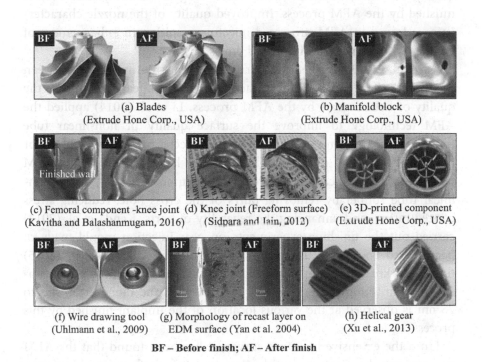

(a) Blades
(Extrude Hone Corp., USA)

(b) Manifold block
(Extrude Hone Corp., USA)

(c) Femoral component -knee joint (Kavitha and Balashanmugam, 2016)

(d) Knee joint (Freeform surface) (Sidpara and Jain, 2012)

(e) 3D-printed component (Extrude Hone Corp., USA)

(f) Wire drawing tool (Uhlmann et al., 2009)

(g) Morphology of recast layer on EDM surface (Yan et al. 2004)

(h) Helical gear (Xu et al., 2013)

BF – Before finish; AF – After finish

FIGURE 3.7 Applications of AFM process.

technique, they have achieved the required level of accuracy and surface finish that can not be met by the conventional method. The various applications of the AFM process are presented in this section.

Perry and Stackhouse (1989) identified the applications of AFM process in the finishing of gas turbine components, namely axial rotor, centrifugal rotors, stators, turbine blades, compressors, turbine disks, and rotating shafts. Further, Kim and Kim (2004) effectively applied AFM technology for deburring of burrs in spring collect made of chrome-molybdenum material. They have reported that the velocity of abrasive media affect the deburring of burrs in spring collects. Yin et al. (2004) developed the abrasive flow polishing machine with turbulence flow characteristics for polishing of S45C steel bores of ⌀400 μm and ⌀500 μm, stainless steel 304 bores of ⌀500 μm, and zirconia bores of ⌀260 μm and achieved 60% improvement in surface finish. Tzeng et al. (2007b) finished micro-slit width of 0.23 mm using AFM and produced slit on stainless steel plate using wire-EDM and found that this process enhances the finish of the micro-slit. Jung et al. (2008) studied the quality of direct injection (DI) diesel engine fuel injector nozzles finished by the AFM process. Improved quality of the nozzle characteristics is found in AFM-processed injectors resulting in enhancement of engine performance and improved emissions.

Xu et al. (2013) finished the burrs present at the interaction between the tooth surface and the end surface of the helical gears to improve the quality of helical gears by the AFM process. Li et al. (2014) applied the AFM technology to improve the surface quality of non-linear tube runner, which is commonly used in some special passage exits of major parts in the field of military and civil. They have concluded, the AFM technology is significant in improving the surface integrity of the non-linear tube runner by removing burrs of the cross hole, reducing stress concentration, and enhancing the reliability of parts. Kenda et al. (2014) carried out the finishing experiments on gear injection mould made of heat-treated tool steel using the AFM process. Kumar et al. (2015) finished a knee joint implant using rotational-magnetorheological abrasive flow finishing process. Surface roughness in the range of 35 nm to 78 nm is observed at the various location of the implant finished by this process.

From the extensive literature survey, it has been found that the AFM process can be used to finish any type of surfaces made of any engineering materials with reduced production time without disturbing the geometry. It should be noted that the surfaces finished with any type of finishing process have an effect on the functionality of the product when it is put into the service. Especially the biomaterials surface qualities have an influence on the surrounding environment. Hence, in the next section, the effect of roughness on the wettability of the surfaces and bacterial adhesion is explained with the available literature.

From the detailed study on the implant machining and finishing, it is understood that most of the implants have a complex and partly free-form surface, which raises the difficulty in finishing in a single step. Also, one should have the knowledge of frequently changing contact conditions while performing the finishing operation to avoid the non-uniform surface quality. This influences the bacterial adhesion and further forms the biofilms on the implants. This calls for the development of advanced techniques that can finish these freeform surfaces with less manufacturing cost and time without damaging the implant's geometries.

Further, in the next section, motivation for taking the present research work is detailed based on the detailed literature survey on the different aspects of AFM process and influence of the finishing on the wettability and bacterial adhesion.

3.3 MOTIVATION

The functionality of the mating components greatly depends on the surface finish and lay patterns that play a significant role in wear resistance, wettability, bacterial adhesion, corrosion resistance, assembly toler-ances, and contact rigidity. These properties are very much essential in hydraulic components, bio-implants, aerospace components, and auto-motive components. Most of these components have complex internal and external surfaces that are difficult to finish with conventional finishing processes due to the restriction for the tool to reach these complex surfaces.

The surface finish of the surgical instruments and implants are kept top priorities in the biomedical industry. Despite the adoption of advanced technology in manufacturing the implants and in surgical and medical management procedures, there are still a large number of implants that are subject to failure due to the surface finish achieved on the implant surfaces. The major causes of implants' failure are fracture, prosthetic dislocation, loosening, excessive wear rate at mating surfaces and its associated debris, and pre-surgical contamination/infection (i.e., bacterial adhesion). Surfaces roughness is one of the factors which influence the bacterial adhesion on implants and acts as favourable sites for colonization and biofilm formation leads to the prosthetic implant infections (PIIs). Most of the implants have a complex and partly freeform surface that raises the difficulty in finishing at a single step. Hence, there is a need for a process that can finish complex internal and external surfaces with better surface finish and lay patterns without any difficulty and economically. This is the main reason for the current investigation.

3.4 OBJECTIVE AND SCOPE OF PRESENT WORK

To overcome the challenges associated with conventional finishing processes to finish complex internal and external surfaces and biomaterials having free form surfaces, a novel unidirectional abrasive flow finishing (UAFF) process is developed and studied the effect of critical process parameters in achieving required surface roughness. The essential parameters of interest are the number of cycles, pressure, and abrasive media that can be varied at different levels to study their effect on output responses-average surface roughness, contact angles, and bacterial adhesion.

3.4.1 The Detailed Scope of Work-Activities Includes

1. Development and fabrication of UAFF experimental setup with various subsystems.

2. Development and characterization of viscoelastic polymer abrasive media.

3. Preliminary experiments on machinability study to understand the effect of the number of cycles on finishing of different materials.

4. The effect of surface roughness and surface morphology of finished biomaterials wettability study.

5. The effect of surface roughness and surface morphology of finished biomaterial surfaces on the bacterial adhesion study.

3.5 METHODOLOGY OF RESEARCH WORK

The work started with a comprehensive literature survey on the different variants of the abrasive flow machining process and the effect of surface roughness on the wettability and the bacterial adhesion. The research work is divided into three main sections, namely, development of experimental setup, development of abrasive media, and experimentation. The experimental setup has been developed in two stages – unidirectional abrasive flow machining and closed loop unidirectional abrasive flow machining. The main components along with subsystems, are selected based on the arrived design parameters. Simultaneously, viscos-elastic polymer-based abrasive media has been developed with varying sizes of abrasive mesh size for the finishing of surfaces. In the present investigation, three sets of experiments are carried out: (1) machinability study has been carried out on the different workpiece materials having different hardness values to study the effect of the number of cycles on the surface roughness, (2) wettability study on the machined biomaterials to investigate the influence of process parameters on surface roughness and contact angle, (3) bacterial adhesion study on the machined biomaterials to investigate the influence of process parameters on surface roughness and bacterial adhesion. The results are presented under different chapters.

3.6 SUMMARY

The importance of surface finish, different surface characteristics, and available surface finishing processes are explained in detail. The material removal mechanism and various types of abrasive flow machining processes are studied. The collected literature on abrasive flow machining is explained under six different sections. Further, the motivation, objectives, and scope of the present work, methodology, and organization of the thesis are presented. The various work-activities carried out in the development of the experimental setup are explained in the next chapter.

CHAPTER 4

A Novel Approach for Finishing Various Implants

UAFF Process

4.1 INTRODUCTION

Freeform surfaces and components with complex internal cavities find wide applications in biomedical, micro-electro-mechanical systems, micro-fluidics, and automotive industries. Finishing of these surfaces to the required level is challenging due to the varying surface and inaccessible internal features. In the present work, a lab scale unidirectional abrasive flow finishing (UAFF) experimental setup is developed and fabricated to finish the complex internal and external features efficiently and effectively. This process can efficiently finish the surfaces without damaging the surface features due to the flexibility of the abrasive media that has the ability to take the shape of the component during the finishing process. The major challenges in the development of the experimental setup are:

- Development of flexible abrasive media

- Selection of proper materials for abrasive media cylinder and hydraulic cylinders for the long run

- Selection of hydraulic power pack that is capable of pushing the highly viscous abrasive media

- Design of hydraulic circuit for switching the actuators automatically

- Development of fixtures to hold different workpieces of different sizes and shapes rigidly during the machining operation

The main objective of this work is to develop the unidirectional abrasive flow concept, that is, abrasive media flows only in one direction to achieve the uniform lay patterns on the work surface in one direction only, which finds wide application in lubrication, biomedical, and surface engineering.

The experimental setup mainly consists of the hydraulic power pack, a hydraulic cylinder, an abrasive media cylinder, direction control valves, and pressure gauges. The development of the experimental setup has been carried out in two stages: Stage 1: UAFF process and Stage 2: closed loop UAFF (CLUAFF) process. The detailed design parameters and fabrication details of the experimental setup are elaborated in this chapter.

4.2 DEVELOPMENT OF EXPERIMENTAL SETUP

The main requirement in the development of experimental setup is to push viscoelastic polymer abrasive media, which is semisolid in nature, through the workpiece efficiently and effectively. The experimental setup should be capable of pushing this high viscous abrasive media. This section elaborates on the design features, and fabrication of the experimental setup carried out at different stages.

4.2.1 Stage 1: Unidirectional Abrasive Flow Finishing Process

Figure 4.1 (a) shows the schematic diagram of the UAFF experimental setup. It consists of an abrasive media cylinder coupled to the hydraulic cylinder, 4/3 solenoid operated directional control valve, pressure gauges, hydraulic power pack, and flexible hoses. The hydraulic cylinder is driven through the hydraulic power pack to push the abrasive media present in the abrasive media cylinder through the workpiece fixed rigidly at the exit using the developed fixture. Pressure gauges are fixed at the hydraulic cylinder to monitor the variation in hydraulic pressure at which the media is pushed through the workpiece. During the

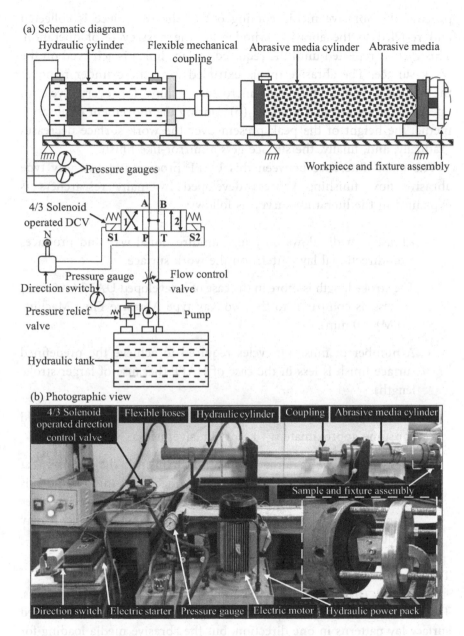

FIGURE 4.1 Unidirectional abrasive flow finishing process.

process, the abrasive media coming out of the workpiece is collected and re-filled to the abrasive cylinder for the next cycle of operation. This cycle is repeated until the required surface finish is achieved on the work surface. The abrasive media extruded from the cylinder removes the material in the form of micro-/nano-chips for every cycle of operation, as shown in Figure 4.1 (b). During the movement of abrasive media, the height of the peak present over the work surface decreases gradually, and, finally, the surface gets a mirror-like finish.

The main difference between this UAFF process and two-way type abrasive flow finishing process developed by many researchers is explained in the literature survey as follows:

- Abrasive media flows only in a unidirectional way and produces a unidirectional lay pattern on the work surface.

- The stroke length is more in the case of a developed UAFF (400 mm) process as compared to the two-way type Abrasive Flow Machine (AFM) (70 mm).

- A number of finishing cycles required to achieve the predefined surface finish is less in the case of UAFF (because of larger stroke length).

- In UAFF, one complete stroke of the abrasive cylinder would consume approximately 4 kg of abrasive media.

- Even though the material removal mechanism is the same as that of the two-way type abrasive flow machine, this method is different in the sense that, in a two-way type, media reciprocates to and fro inside the workpiece with small cylinder stroke. It may produce criss-cross lay patterns over the work surfaces.

Table 4.1 shows the specifications of the UAFF experimental setup.

Figure 4.1 (b) shows the photographic view of the UAFF process. This setup has the capability of finishing any features with controlled surface lay patterns in one direction, but the abrasive media loading for the next cycle is difficult. This should be carried out manually and, hence, it is time-consuming during the loading process. To avoid the manual reloading and to achieve the automatic recirculation of the abrasive media closed-loop concept, an experimental setup has been developed and is explained in Section 4.2.2.

TABLE 4.1   Specification of the unidirectional abrasive flow finishing experimental setup

Components	Specifications
Hydraulic cylinder	Cylinder: ⌀63 mm
	Piston: ⌀36 mm
	Stroke length: 500 mm
	Pressure: 250 bar
Abrasive media cylinder	Cylinder: ID ⌀100 mm, OD ⌀110 mm
	Piston: ⌀100 mm
	Length: 500 mm
	Material: EN8 (Honed)
Hydraulic power pack	Motor: Electric motor
	Pump: Gear pump
	Power: 1.2 kW
	Pump capacity: 17 lt/min
	Working pressure: 100 bar
	Hydraulic fluid: Biodegradable oil
	Tank capacity: 100 lt
Control valve	Type: 4/3 Solenoid operated directional control valve
	Make: YUKEN, Germany
	Type: Open centre
	Pressure capacity: 309 bar

4.2.2 Stage 2: Closed Loop Unidirectional Abrasive Flow Finishing Process

To overcome the difficulty in loading the abrasive media for the next cycle, a simple, compact, and economical hydraulic powered CLUAFF setup has been developed, as shown in Figure 4.2. The setup consists of two hydraulic cylinders, two abrasive media cylinders; two 4/3 solenoid operated directional control valves, hydraulic power pack, and accessories like two pressure gauges, four limit switches, flexible hoses, and two work holding fixtures. Each hydraulic cylinder is coupled with abrasive media cylinder using the flexible mechanical coupling to transmit positive linear motion. The hydraulic power pack of capacity 100 bar is used to drive the hydraulic cylinders, and these are connected through flexible hoses. Two 4/3 solenoid operated directional control valves, namely DCV 1 and DCV

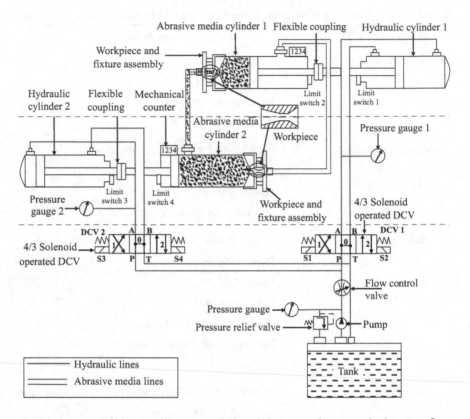

FIGURE. 4.2 Schematic diagram of closed loop unidirectional abrasive flow finishing setup.

2 are used to control the direction of hydraulic oil flow to change the direction of the abrasive media cylinder. The extreme positions of the hydraulic cylinders are monitored using four limit switches: limit switch 1 and 2 used for hydraulic cylinder 1; and limit switch 3 and 4 used for hydraulic cylinder 2. The working pressure of the hydraulic cylinder is monitored using two pressure gauges fixed at the inlet of hydraulic cylinders 1 and 2. Two fixtures are developed and placed at the exit of the abrasive media cylinder 1 and 2 to hold the workpieces rigidly at the proper position.

Table 4.2 shows the sequence of operations used to achieve the closed loop Unidirectional Abrasive Flow Machining (UAFM). Hydraulic cylinder 1 extends from limit switch position 1 to limit switch position 2. During its travel, it pushes the media present in abrasive media

TABLE 4.2   Sequence of operation

Direction control valve	Valve position	Solenoid position	Operation
DCV 1	0	S1 and S2 OFF	No cylinder movement. Hydraulic cylinder 1 and 2 are at position limit switch 1 and limit switch 3
	1	S1 ON	Hydraulic cylinder 1 extends and thereby pushing the abrasive media through abrasive media cylinder 1 until it reaches limit switch 2 and the pushed abrasive media falls at abrasive media cylinder 2.
	2	S2 ON	Hydraulic cylinder 1 retract and thereby abrasive media cylinder 1 also retract
DCV 2	0	S3 and S4 OFF	No cylinder movement. Hydraulic cylinder 1 and 2 are at position Limit Switch 1 and Limit Switch 3
	1	S3 ON	Hydraulic cylinder 2 extends and thereby pushing the abrasive media through abrasive media cylinder 2 until it reaches the Limit Switch 4, and the pushed abrasive media to pass through a flexible hose to abrasive media cylinder 1.
	2	S4 ON	Hydraulic cylinder 2 retracts, and thereby abrasive media cylinder 2 also retracts.

cylinder 1 through the workpiece fixture 1 mounted at the exit of abrasive media cylinder 1. The extruded abrasive media falls inside the abrasive media cylinder 2 through the gravity feed. After one complete forward stroke of cylinder 1 (limit switch position 1 to limit switch position 2), DCV 1 receives the control signal to divert the flow back to the hydraulic cylinder 1 to move from limit switch position 2 to limit switch position 1 and bring back the abrasive media cylinder 1. Once the hydraulic cylinder 1 retracted back to the original position (limit switch 1 position), DCV 2 receives the control signal to send the fluid flow to the hydraulic cylinder 2.

Then hydraulic cylinder 2 extends from limit switch position 3 to limit switch position 4 and it pushes the abrasive media present in the abrasive media cylinder 2 through workpiece fixture 2 mounted at the exit of the abrasive cylinder 2. Now the extruded abrasive media from abrasive media cylinder 2 is sent back to the abrasive media cylinder 1 through the flexible hose. After one complete forward stroke of abrasive media cylinder 2 (limit switch position 3 to limit switch position 4), DCV 2 receives the control signal to divert the flow to bring back hydraulic cylinder 2 from limit switch position 4 to limit switch position 3 and in turn it bring back the abrasive media cylinder 2 to the original position, and this completes the one closed loop cycle. This cycle will repeat automatically to re-circulate the media continuously through the workpieces until the required finish is achieved.

4.2.2.1 Design Parameters

The theoretical calculations are carried out using the basic equations to decide the size of the basic components. Based on the calculations, available standard size components are selected, and some of the components are fabricated in-house.

Some of the major requirements while selecting the components and other subsystems are as follows:

- Setup should have the capacity to push the semi-solid flexible abrasive medium

- Closed loop for automatic recirculation of abrasive medium for next cycle

- Provision for clamping multiple workpieces

- Capability to withstand the varying pressure and flow velocity

- Suitable power sources, sensors, and instrumentation for ease of machining

For sizing the various components, the following assumptions have been made:

- The maximum velocity of the abrasive media in the passage is 200 mm/s.

- The internal diameter of the components to be finished varies between 1 mm and 18 mm.

- The minimum time required to complete one pass of 400 mm is 1 min.

The maximum flow rate of the abrasive media is calculated by taking into consideration the maximum media flow velocity and the maximum internal diameter of the workpiece to be finished.

The maximum media flow rate, Q_m, is given by

$$Q_m = \left(\frac{\pi}{4} d_w^2 \times v_{max}\right) \qquad (4.1)$$

where d_w is the maximum internal diameter of the workpiece to be finished, which is 18 mm and v_{max} is the maximum media flow velocity inside the component, which is about 200 m/s.

$$Q_m = \frac{\pi}{4}(18)^2 \times 0.2 = (50.89 \times 10^3) \text{ mm}^3/\text{s}$$

Assuming the minimum time (t) required for one complete cycle as 60 s, then the required volume (V) of the working cylinder is

$$V = (Q_m \times t) = (50.89 \times 10^3 \times 60) = (3.054 \times 10^6) \text{ mm}^3 \qquad (4.2)$$

In order to restrict the stroke length 400 mm, an abrasive media cylinder diameter d_a = 63 mm and stroke length of l_s = 400 mm is selected to meet the requirement.

$$Q_m = A_a v_a \qquad (4.3)$$

The maximum velocity (v_a) of the piston in the abrasive cylinder of area (A_a) is given by

$$v_a = \left(\frac{Q_m}{A_a}\right) = \left(\frac{Q_m}{\frac{\pi}{4} d_a^2}\right) = \left(\frac{50.89 \times 10^3}{\frac{\pi}{4} 63^2}\right) = 16.32 \text{ mm/s} \qquad (4.4)$$

where A_a is the area of the abrasive media cylinder and d_a is the diameter of the abrasive media cylinder.

The operating pressure range of the abrasive flow finishing process in the available literature is in between 10 bar and 100 bar. Hence, in the present work, maximum operating pressure (developed in the abrasive cylinder) under all test conditions is assumed to be P_{max} = 100 bar.

Based on the maximum operating pressure 100 bar and a stroke length of 400 mm, the standard hydraulic actuator is selected. Hydraulic actuator specifications are piston diameter: ⌀63 mm, piston rod: ⌀28 mm, stroke length: 400 mm, and pressure 250 bar.

The working areas of the hydraulic actuator are

$$\text{Forward direction, } A_f = \left(\frac{\pi}{4} \times 63^2\right) = 3,110 \text{ mm}^2 \qquad (4.5)$$

$$\text{Return direction, } A_r = \left(\frac{\pi}{4} \times 28^2\right) = 615 \text{ mm}^2 \qquad (4.6)$$

where A_f is an area of hydraulic cylinder piston end (forward direction) and A_r is an area of hydraulic cylinder rod end (return direction).

Based on the maximum velocity of the piston in the abrasive media cylinder, the required maximum flow rate Q_{max} is given by

$$Q_{max} = A_f \times v_a = 3,110 \times 16.32 = 50.75 \times 10^3 \text{ mm}^3/\text{s} \quad (4.7)$$

The maximum pressure likely to be in the hydraulic cylinder p_h is

$$p_h = p_{max}\left(\frac{A_a}{A_f}\right) = 100 \left(\frac{3,110}{3,110}\right) = 100 \text{ bar} \qquad (4.8)$$

Assuming, the total efficiency $\eta_t = 0.9$, the required power input P at the hydraulic power pack is

$$P = \left(\frac{p_h\, Q_{max}}{\eta_t}\right) = \left(\frac{100 \times 50.75 \times 10^3}{0.9}\right) = 0.45 \text{ kW} \qquad (4.9)$$

Specification of CLUAFF experimental setup is presented in Table 4.3. Details of the individual components selected for the assembly is explained in Section 4.3.

4.3 SUBSYSTEMS

The major components of interest are hydraulic power pack, hydraulic cylinder, abrasive media cylinder, directional control valve, pressure gauges, base support, and hoses. Selected subsystems along with main specifications, are explained in detail in the following text.

4.3.1 Hydraulic Power Pack

The abrasive media used to finish the work surfaces is semi-solid flexible viscoelastic polymer media. To push this media through the workpiece,

TABLE 4.3   Specification of closed loop UAFF experimental setup

Components	Required no.	Specifications
Hydraulic cylinder	2	Cylinder: ⌀63 mm
		Piston: ⌀28 mm
		Length: 400 mm
		Pressure: 250 bar
Abrasive media cylinder	2	Cylinder: ID ⌀63 mm, OD ⌀75 mm
		Piston: ⌀63 mm
		Length: 430 mm
		Material: EN8 (Honed)
Hydraulic power pack	1	Motor: Electric motor
		Pump: Gear pump
		Power: 1.2 kW
		Pump capacity: 17 lt/min
		Working pressure: 100 bar
		Hydraulic fluid: Biodegradable oil
		Tank capacity: 100 lt
Control valves	2	4/3 Solenoid operated directional control valve
		Make: YUKEN, Germany
		Type: Open centre
		Pressure capacity: 309 bar

a high pressure hydraulic power pack is required. So, the experimental setup is connected to the hydraulic power pack, which has the capacity of pumping fluid at 17 lt/min at 100 bar pressure. Figure 4.3 shows the circuit diagram and the photographic view of the hydraulic power pack. It consists of a reservoir, filters, pump, electric motor, pressure relief valve, and electric stator to control the motor. The specifications of the hydraulic power pack used in the present work are listed in Table 4.4. The capacity of the pump is 17 lt/min at a maximum pressure of 100 bar. The reservoir is having the capacity of storing 100 lt of hydraulic oil. The suction line is attached with pressure line filter that is connected to the pump, and in the same line, two outlet port with quick connect coupling is provided for possible connection of two actuators at one time. The pump is driven by the three-phase induction motor, which is connected

(a) Circuit diagram

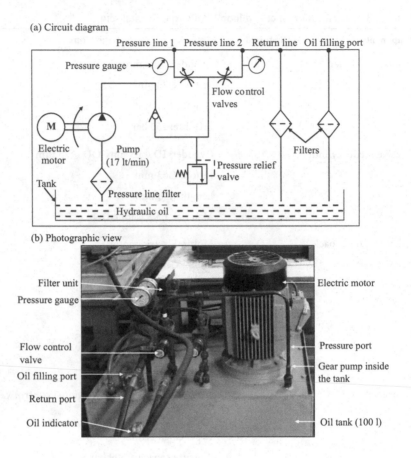

(b) Photographic view

FIGURE 4.3   Hydraulic power pack.

TABLE 4.4   Specification of the hydraulic power pack

Make	Assembled
Operating pressure	100 bar
Flow rate	17 lt/min
Tank capacity	100 l
Electric motor	3-phase induction motor
Pump	Gear pump

to the electric stator. One return line is connected to the tank with the filter unit. Two pressure lines are connected to the two pressure gauges for measuring the pressure at the pressure line.

4.3.2 Hydraulic Cylinder

Figure 4.4 (a) shows the schematic diagram of the hydraulic cylinder used in the present work. Based on the calculations and availability, a standard size double acting hydraulic cylinder of HT-63-28-CC-MF3 -400 is selected. The selected hydraulic cylinder is of 63 mm bore diameter and 400 mm stroke length. The other specifications of the hydraulic cylinder are given in Table 4.5.

4.3.3 Abrasive Media Cylinder

The abrasive media is filled inside the abrasive media cylinder, and the hydraulic power is used to push the abrasive media through the work-piece. For the successful running of the experimental setup for a long time, material selection and fabrication of abrasive media cylinder is very important. It has to be hard enough to resist the scratches from abrasives and also super finished for easy flow of media without friction. In the present work, the abrasive cylinder of ⌀75 mm (OD) × ⌀63 mm (ID) × 430 mm (length) made of EN8 material is used, and the internal surface is honed to remove the irregularities.

4.3.4 Directional Control Valve with the Manifold Block

In the present work, a 4/3 solenoid operated directional control valve (DCV) is used in the experimental setup for close control over the hydraulic cylinder operation. Figure 4.4 (b) shows the 4/3 solenoid operated directional control valve. The detailed specifications of the 4/3 solenoid operated direction control valve are listed in Table 4.6.

4.3.5 Pressure Gauges

Extrusion pressure is one of the main parameters that affect the finishing process. To monitor the extrusion pressure, pressure gauges are mounted in both forward and return line of the hydraulic cylinders. The symbol and photographic view of the pressure gauge are shown in Figure 4.4 (c). The pressure gauge can read the pressure between (0–275) bars, and further specifications are given in Table 4.7.

(a) Hydraulic cylinder

Geometry   Symbol

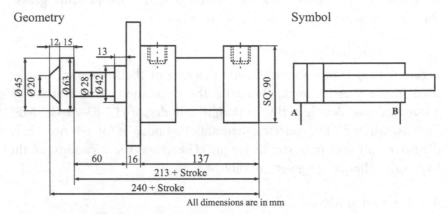

All dimensions are in mm

(b) Solenoid operated directional control valve

Photographic view   Symbol

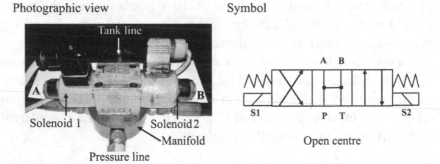

Open centre

(c) Pressure gauge

Photographic view   Symbol

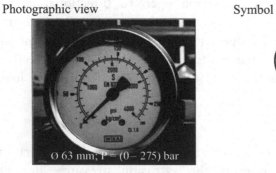

FIGURE 4.4   Subsystems of the experimental setup.

TABLE 4.5 Specification of hydraulic cylinder

Make	VELJAN, India
Model specification	HT - 63 - 28 - CC - MF3 - 400

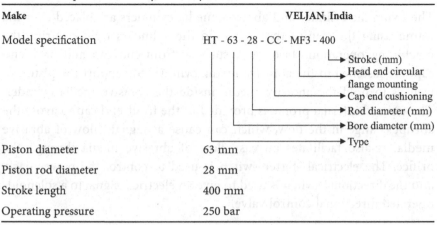

Piston diameter	63 mm
Piston rod diameter	28 mm
Stroke length	400 mm
Operating pressure	250 bar

TABLE 4.6 Specification of 4/3 solenoid operated direction control valve

Make	YUKEN, Japan
Model	DSGR-01-3C3-A240-2080
Number of port	4
Number of position	3
Operating pressure	309 bar
Tank line back pressure	156 bar
Flow rate	63 lt/min
Max. change over frequency	300 cycles/min
Mass	2.2 kg

TABLE 4.7 Specification of pressure gauge

Make	WIKA, Germany
Model	213.53.63 Bourdon Tube (Glycerine: Liquid Filling)
Operating pressure	0 to 275 bar
Dial size	63 mm
Connection	Bottom 1/4″ BSPM
Working Temperature	– 20 to + 60°C

4.3.6 Other Accessories

The hydraulic cylinder and abrasive media cylinders are placed over the frame using the clamps that can hold the cylinders rigidly during the machining operation. Two caps, such as front-end cap and back-end cap, are fastened to the abrasive media cylinder to support the piston as well as to hold the abrasive media inside the abrasive media cylinder. The tapered internal profile is provided in the front-end cap to avoid the abrupt change in the flow, which can cause a stagnant flow of abrasive media. It also facilitates an easy flow of abrasive media through the orifice. The electrical starter switch is used to control the power pack, and the directional switch is used to give an electrical signal to a solenoid-operated directional control valve.

4.4 WORKPIECE AND FIXTURE DESIGN

In the present work, experimentations have been carried out with different objectives to explore the flexibility of the developed process in meeting the industrial standards for finishing various materials and features. To carry out these experiments on different workpiece materials with different shapes, different workpiece fixtures are developed and fabricated to hold the workpieces rigidly during the machining operations, as detailed in the following sections.

4.4.1 Fixture to Hold the Workpieces for Machinability Study

The main objective of this experiment is to study the effect of process parameters on the machinability and finish achieved on four different engineering materials and to understand the material behaviour under the finishing parameter. For the machinability study, aluminium, brass, copper, and mild steel workpieces are selected to study the effect of the finishing process. These materials are commonly used in almost all industrial applications such as automotive, aerospace, electrical, biomedical, and naval industries. To hold the workpieces, a special type of stainless steel fixture is developed and fabricated, which can hold four workpieces at a time. The workpiece and fixture geometry and dimensions are shown in Figure 4.5 (a). Two end caps and supporting holders are attached with the fixture to restrict the movement of the workpieces during the media flow.

(a) Photographic view of workpieces and fixture assembly used for the machinability study

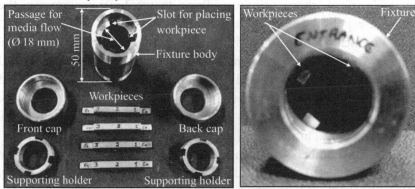

(b) Workpiece and fixture assembly used for wettability study

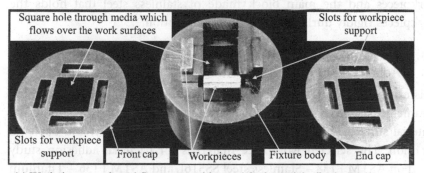

(c) Workpiece sample and fixture assembly used for bacterial adhesion study

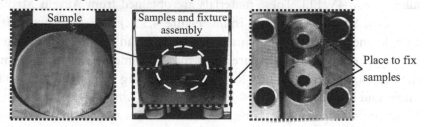

FIGURE 4.5   Workpieces and fixtures.

4.4.2 Fixture to Hold Biomaterials Used for Wettability Study

The objective of this study is to finish biomaterials and to understand the effect of process parameters like the number of cycles and pressure on the surface roughness, surface topography, and how these finished surfaces influence the contact angle and surface energy. Biomaterials SS316L and Ti-6Al-4V are selected for the study. The materials are received in bulk form. For experimentation purposes, the samples are prepared based on the required size (20 mm × 10 mm × 6 mm) using a wire electric discharge machine. Prior to the finishing process, the samples are polished using metallographic SiC papers having the grit size 220 (8-inch diameter) placed in the polishing machine for 120 s time duration, with a speed of 150 rpm. Later the samples are cleaned with DI water, ethanol, and ultrasonicated to remove the polished residue. To avoid difficulties in measuring the surface roughness and contact angle, flat work surfaces with small size samples are considered for the experimentation, instead of complex, intricate cavities. The fixture can hold multiple workpieces – four numbers and hence finishes the four workpieces in one stretch. Figure 4.5 (b) shows the developed workpiece and fixture assembly used for wettability study. It consists of two end caps that support the workpieces and the main block made of stainless steel that holds the workpieces rigidly during the machining operation.

4.4.3 Fixture to Hold the Biomaterials Used for Bacterial Adhesion Study

This experimental study aims at understanding the influence of finishing parameters such as a number of cycles and an abrasive particle size on the material surface roughness, surface topography, wettability, and bacterial adhesion. The selected workpiece materials for bacterial adhesion study based on ASTM F138 – Stainless steel SS316L and ASTM F136 – Titanium alloy Ti-6Al-4V ELI. These materials are obtained from Auxein Medical Pvt. Ltd., Sonepat, Haryana, India. Figure 4.5 (c) shows the workpiece sample and fixture assembly used for bacterial adhesion study. Circular disc type of sample is prepared for the experimentation with 13 mm diameter and 6 mm thickness for ease of measuring the roughness, contact angle, and bacterial adhesion study. To hold the workpieces, stainless steel fixture is developed, which consists of two blocks with two circular slots on each block for placing the samples. These two blocks are assembled tightly with screws, and also work samples are screwed to the fixture at one end to avoid the movement during the finishing operation.

4.5 MATERIALS AND ABRASIVE MEDIA DEVELOPMENT

The abrasive polymer media used in the present investigation is a mixture of viscoelastic polymer, plasticizer, and suitable abrasive particles. Extrude Hone Corporation, USA, and Kennametal, USA is the commercial manufacturers of the abrasive media, but it is very costly, and they have not revealed any constituents of the abrasive media. To overcome this, few researchers have made an attempt to develop their own abrasive media and studied the media properties, which are discussed in a literature survey of Chapter 3. From the literature, it is clear that the development of abrasive media is a challenging task. So the abrasive media should possess the following properties to meet the standards and general customer requirements as per the literature survey:

- The recommended minimum viscosity of abrasive media is 30 kPa.s to achieve a better finish on complex surfaces (Petare and Jain, 2018).

- It should be self-deformable eventually find its own shape and way when placed in a container with a given time. This property also helps in finishing the complex internal cavities and the external surfaces because the abrasive media has the ability to take the shape of the cavities.

- It should have the property of non-sticking and should be chemically inactive and non-corrosive in nature.

- It should have the good abrading ability and provide sufficient force on the abrasive particles over the work surface during the finishing process.

Hence, the appropriate selection and composition of the individual constituents play a major role in the better binding of polymer and abrasive particles. In this section, the detailed report on the selection of individual materials, preparation of media, and characterization of the abrasive media are detailed.

4.5.1 Viscoelastic Polymer

A viscoelastic polymer commonly used in the development of abrasive media acts as a carrier medium and carries abrasive particles. It has viscous and elastic properties that help in distributing the axial and radial force on the abrasive particles with the help of applied extrusion pressure. Some of the commonly used polymer carriers are polyborosiloxane, silicone rubber, silly putty, natural rubber, styrene butadiene rubber, butyl rubber, etc. In the present investigation, silicone rubber is selected as a viscoelastic polymer media (carrier media) because of its unique properties like wide temperature range (−101°C to 316°C), less cost compared to other synthetic rubbers (Aggarwal, 1987), good thermal conductivity, abrasion resistance, chemical stability (Shit and Shah, 2013), high tear strength and elongation, good processability and not sticky to surfaces, which are the basic requirements for abrasive media. Also, it is one of the most widely used in manufacturing components of aerospace, aviation, medical, semiconductor, and automotive industries. The literature shows that silicone rubber is the best suitable abrasive media, and it has compatibility with a wide range of abrasive particles and is also mechanically stable. The selected type of silicone rubber is Bluesil HCR 1940 LA2, and its physical properties are listed in Table 4.8.

4.5.2 Plasticizer

The plasticizer is also called processing oil, and they are small molecules (low molecular weight) and posses small chains. These are chemically similar to the polymers, diffuses between high molecular weight polymer chains that increase the gap between the polymer chains by providing greater mobility and converts solid to semisolid (Sankar et al., 2011).

Some of the commonly used processing oils are naphthenic oil, varnish oil, silicone oil, etc. In the present research work, silicone oil is selected as a plasticizer as because it has better mixing compatibility with silicone-based polymers. After adding the plasticizer, the base viscoelastic polymer attains better flowability, self deformability, and

TABLE 4.8 Physical properties of silicone rubber

Make	Bluestar Silicones, China
Grade	Bluesil HCR 1940 L A2
Appearance	Translucent
Density	1,120 kg/m^3
Shore A hardness	42
Tensile strength	9.2 MPa
Viscosity	21,000 Pa.s
Tear strength	2,200 kN/m
Rebound resilience	58%

flexibility to reach any restricted openings present in the workpiece. Table 4.9 shows the physical properties of the plasticizer.

4.5.3 Abrasive Particles

Abrasive particles are the basic constituents of the abrasive media and they act as a cutting tool to remove the material from the work surfaces. The basic requirements for the abrasive particles are: (a) should be harder than workpiece to penetrate on the work surface and (b) compatible with the polymer carrier media and the plasticizers for better binding (Kar et al., 2009a). In the present work, silicon carbide (SiC) is selected as the abrasive particle because it is the most commonly used abrasive in the grinding, polishing, and lapping to finish the wide range of workpiece materials. Table 4.10 shows the physical properties of the selected abrasive particle.

TABLE 4.9   Physical properties of plasticizer

Make	Wacker Chemie India Pvt.Ltd., India
Grade	WACKER AK 1000
Appearance	colourless, clear
Density	approx. 970 kg/m^3
Refractive index	1,404
Flash point	> 300°C
Surface tension	0.021 N/m
Ignition temperature (liquid)	410°C
Viscosity	1 Pa.s

TABLE 4.10   Physical properties of abrasive particle

Make	SNAM Abrasives, India
Grade	Mesh size #120 to #800
Appearance	Black and Green colour
Density	3,100 kg/m^3
Hardness (Knoop)	2.63×10^9 kg/m^2
Thermal conductivity	120 W/m °K
Elastic modulus	410×10^9 Pa

4.5.4 Preparation of Abrasive Media

Figure 4.6 (a) shows the steps followed in the preparation of the abrasive media. Silicone rubber polymer media, along with the processing oil – silicone oil and silicon carbide abrasive particles of different mesh size are mixed in definite proportion to prepare the abrasive media. The 50 wt% of SiC abrasive particles, 38 wt% of silicone rubber, and 12 wt% of silicone oil are mixed thoroughly using the two roll mill machine. Initially, the silicone rubber is fed to the roller, and the silicone oil is then added to the polymer to reduce the viscosity and hence increase the flowability. After 5 min of continuous rolling, the selected wt% of SiC particles are added slowly to the mixture and rolled for about another 20 min to achieve a required uniform distribution of SiC in silicone rubber and silicone oil mixture. During the mixing process, the temperature is raised between 40°C and50°C due to the friction between the rollers, and this helps in the continuous breaking of the polymer chain. The nip gap (the gap between two rollers) of 1 mm–0.5 mm is maintained, and the mixing time of 25 min is followed during the media preparation. Figure 4.6 (b) shows the photographic view of the two roll mill used during the preparation of the abrasive media.

The weight of each constituent is arrived using the following equations – Eq. (4.10) to Eq. (4.11).

The volume of the abrasive media cylinder is

$$v_a = \frac{\pi d_a^2}{4} \times L_a \qquad (4.10)$$

where v_a – the volume of the abrasive cylinder (m^3), L_a – length of the abrasive cylinder (m), and d_a – diameter of abrasive media cylinder (m).

The density of the abrasive media

$$\rho_m = (w_a \times \rho_a) + \left(w_p \times \rho_p\right) + (w_o \times \rho_o) \qquad (4.11)$$

where ρ_m – density of abrasive media (kg/m^3), ρ_a, ρ_p, ρ_o – density of abrasive particles, density of polymer media, and density of processing oil, w_a, w_p, w_o – weight percentage of abrasive particles, wt% of polymer media, and wt% of processing oil, respectively.

(a) Preparation steps in the abrasive media

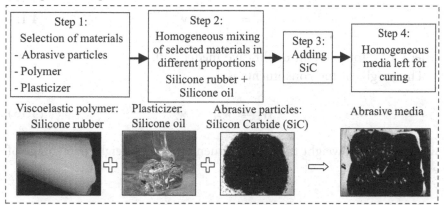

(b) Photographic view of two roll mill machine

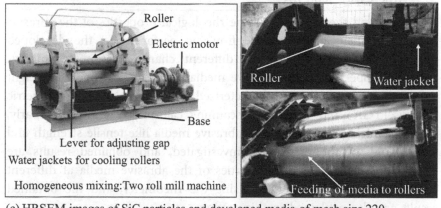

(c) HRSEM images of SiC particles and developed media of mesh size 220

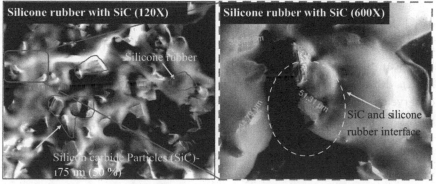

FIGURE 4.6   Preparation of abrasive media.

The weight of the abrasive media is calculated using

$$w_m = \rho_m \times v_a \qquad (4.12)$$

where w_m _ total weight of abrasive media (kg).

The weight of the constituents is

$$W_c = w_m \times w_c \qquad (4.13)$$

where W_c – *the* weight of the constituent and w_c _ weight percentage of each constituent.

4.6 CHARACTERIZATION OF THE DEVELOPED ABRASIVE MEDIA

To understand the thermal and the rheological properties of the abrasive media and their effects on the finishing performance, the developed abrasive media is subjected to different characterization techniques. Thermal properties of the abrasive media like heat flow and the weight loss with temperature are characterized using the thermo gravimetric analysis (TGA) and differential scanning calorimetry (DSC). Similarly, the mechanical properties of the abrasive media like tensile strength and morphology of the media are investigated, and obtained results are presented. The rheological properties of the abrasive media at different temperatures are characterized using the rheometer, and the obtained results are also explained in this section.

4.6.1 Morphology

Morphology of the abrasive particles is observed using the High Resolution Scanning Electron Microscope (HRSEM: Model Inspect F50, FEI, USA). Morphology of particles and silicone rubber interaction is observed using SEM Quanta 400 inspect the instrument. Figure 4.6 (c) shows the HRSEM images of SiC particles and the developed abrasive media of mesh size 220.

It can be seen from the SEM images that the interaction between SiC particles and silicone rubber is good and shows the uniform distribution of particles except for a few local agglomerations of particles. The particles tend to touch each other at most of the places; this may increase the resistance of the media to the conductance of the heat

(Zhou et al., 2007; Kemaloglu et al., 2010). The silicone rubber acts as a carrier medium that carries the SiC particles and plasticizer when it is pushed over the work surface to be finished.

4.6.2 Thermo Gravimetric Analysis (TGA)

Thermal properties of the abrasive media have a significant influence on the viscosity and, in turn, on the finishing process. It is reported in the literature that, during the finishing process, the temperature may rise up to 70°C. This rise in temperature is due to the continuous recirculation of the same abrasive media (Petare and Jain, 2018) and friction between the abrasive media and the work surface. Hence, it is essential to study the thermal properties of the developed abrasive media under different temperatures. TGA is carried out for measurement of decomposition temperature of the abrasive media, which also gives the information on weight loss using a TGA instrument shown in Figure 4.7 (a) (Model SDT Q600, TA Instruments, New Castle, DE, USA).

Abrasive media of size 6 mg is placed on a platinum sample cup and heated from room temperature 29°C to 900°C at 10°C/min under the nitrogen atmosphere at 100 ml/min. The percentage of weight loss to temperature is shown in Figure 4.7 (b), and the changes are observed in three stages. Stage 1: Nearly 5% of weight loss is observed at a temperature of 423°C, and this weight loss is due to moisture present in the abrasive media. Stage 2: At this stage, between 423°C and650°C, maximum weight loss is observed due to the decomposition of silicon rubber and added plasticizer. Stage 3: Above 650°C, very less change in wt% is observed, and at 900°C, the remaining residue is 50%, which is SiC particles. Figure 4.7 (c) shows the maximum degradation temperature is 576°C and the derivative weight loss of 0.424%/°C is observed. Hence, the TGA test helps in finding the minimum temperature that alters the polymer chain that may affect the abrading properties of the abrasive media. From the test, it is clear that the media can perform better at lower temperatures less than 100°C because the weight loss within this temperature is negligible.

4.6.3 Differential Scanning Calorimetry (DSC)

Differential Scanning Calorimetry (Model Q2000 V24 series, TA Instruments, New Castle, DE, USA) shown in Figure 4.8 (a) is used to measure the heat flows and temperature associated with transitions in materials with respect to time and temperature in a controlled atmosphere.

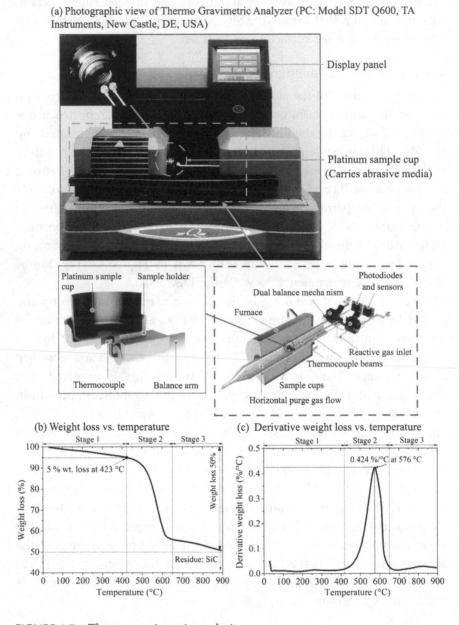

(a) Photographic view of Thermo Gravimetric Analyzer (PC: Model SDT Q600, TA Instruments, New Castle, DE, USA)

— Display panel

— Platinum sample cup (Carries abrasive media)

Platinum sample cup   Sample holder

Thermocouple   Balance arm

Photodiodes and sensors

Dual balance mechanism

Furnace

Reactive gas inlet
Thermocouple beams

Sample cups

Horizontal purge gas flow

(b) Weight loss vs. temperature

Stage 1   Stage 2   Stage 3

5 % wt. loss at 423 °C

Residue: SiC

Weight loss (%)

Weight loss 50%

Temperature (°C)

(c) Derivative weight loss vs. temperature

Stage 1   Stage 2   Stage 3

0.424 %/°C at 576 °C

Derivative weight loss (%/°C)

Temperature (°C)

FIGURE 4.7   Thermogravimetric analysis.

Measurements are conducted at a nitrogen atmosphere (50 ml/min) with varying temperature from room temperature to 500°C at a heating rate of 10°C/min for pure silicone rubber and silicone rubber mixed with abrasive particles (sample size 14 mg). Figure 4.8 (b) shows the DSC curve for

(a) Photographic view of Differential Scanning Calorimetry (PC: Model Q2000 V24 series, TA Instruments, New Castle, DE, USA),

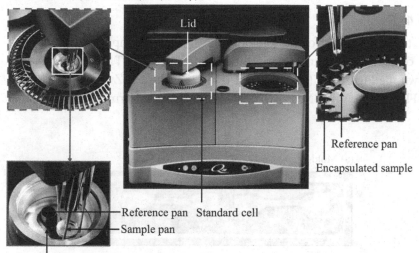

(b) Heat flow vs. Temperature for silicon rubber and abrasive media

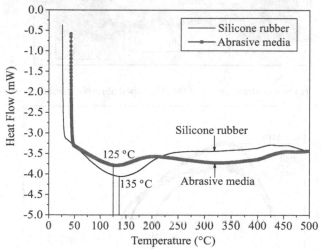

FIGURE 4.8   Differential scanning calorimetry analysis.

silicone rubber and abrasive media. The heat flow curve of the abrasive media is nearly the same as the silicone rubber. The T_g (glass transition temperature) of silicone rubber and abrasive media appears at nearly 125°C and 135°C. The initial decomposition temperature in the curve for silicone rubber is found at low temperatures compared to abrasive media (Zhou et al., 2007).

4.6.4 Tensile Test

The tensile test is performed on a dumbbell-shaped sample that is cut using the compression moulding machine according to ASTM D638 standard shown the Figure 4.9 (a). Universal testing machine – Zwick Roell (Model T1-FRxxMOD.A4K, M/s Zwick GmbH & Co, Germany) is used to

(a) Mold

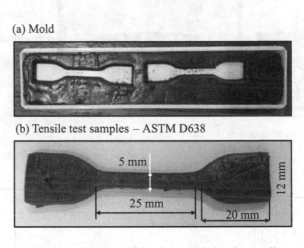

(b) Tensile test samples – ASTM D638

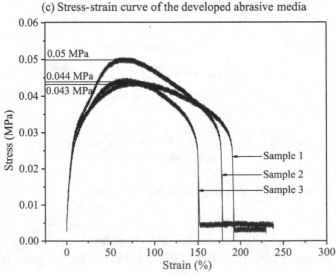

(c) Stress-strain curve of the developed abrasive media

FIGURE 4.9   Tensile test results.

determine the tensile strength of the abrasive media. The test has been performed on three samples with a strain rate of 0.189 s^{-1} at room temperature. Figure 4.9 (c) shows the stress-strain curve of the developed media for three samples. The tensile strength of developed media is 0.045 MPa, strain at break is 175%, and Young's modulus is 1.34 GPa. The tensile strength of the pure silicone rubber is 9.2 MPa as per manufacturer catalogue, which is greater than the developed abrasive media. The addition of 50% of SiC particles decreased the tensile strength of the media because of poor interfacial interaction (Kemaloglu et al., 2010). Further, the addition of the processing oil also affects the properties of the abrasive media.

4.6.5 Rheological Properties

Rheological properties of the abrasive media play an important role in the abrasive finishing process. To understand the media behaviour at different temperatures, rheological properties of the developed abrasive media are ascertained using the rheometer, and results are presented in this section. The static and dynamic rheological properties of the abrasive media such as viscosity, shear stress, storage modulus, loss modulus, and complex viscosity are ascertained using the Anton Paar Physica MCR 301 rheometer subjected to parallel plate configuration. Figure 4.10 shows the schematic diagram and the photographic view of a rheometer used to ascertain the rheological properties of the abrasive media. It consists of a head, in which the main components encoder, motor, and air bearing, are placed that helps in the proper functioning of the rheometer.

The sample of abrasive media of thickness 2 mm and diameter 25 mm is kept between the parallel plates while conducting the experiments. The experiments have been carried out at different temperatures 25°C, 35°C, 45°C, and 55°C, respectively, with varying shear rates from 0.001 s^{-1} to 100 s^{-1} and dynamic frequency sweep test with varying oscillation frequency from 0.001 Hz to 100 Hz. The parameters such as shear rate and frequency ranges are selected based on the available literature (Kar et al., 2009a; Sankar et al., 2011), and maximum extrusion pressure can be varied in the developed experimental setup. Each experiment is repeated for a minimum of three times for repeatability confirmation, and the results are discussed in the following sections.

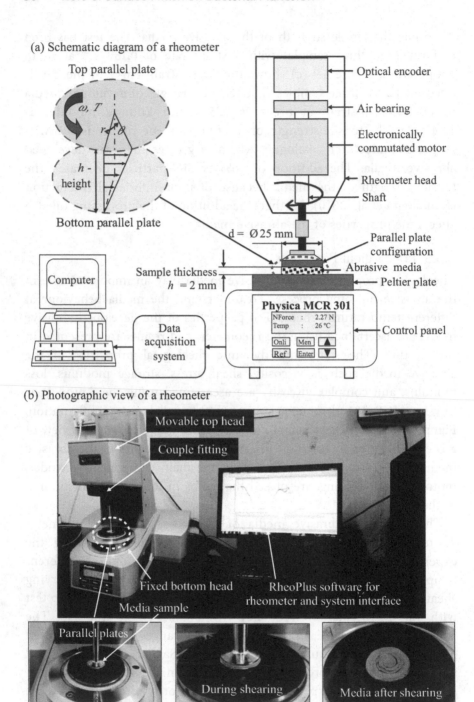

FIGURE 4.10 Rheometer used to ascertain the rheological properties of the abrasive media.

4.6.5.1 Viscosity vs. Shear Rate

The viscosity of the media is an important rheological parameter for describing the flow of abrasive media, which depends on the shear rate, temperature, loading, abrasive size, abrasive shape, and abrasive concentration (Kar et al., 2009a). In the present investigation, the effect of shear rate and the temperature on the viscosity of the abrasive media for 220 mesh size is examined. Figure 4.11 (a) shows the viscosity vs. the shear rate plot at 25°C. The variation of the viscosity with respect to shear rate 0.01 s^{-1} to 100 s^{-1} and is separated into three regions: Region 1 – Newtonian behaviour is observed at a very low shear rate (0.01 s^{-1} to 0.3 s^{-1}). Region 2 – viscosity decreases drastically with increasing shear rate (0.3 s^{-1} to 5 s^{-1}) shows shear thinning behaviour. Region 3 – The viscosity reduction rate is limited at a high shear rate (5 s^{-1} to 100 s^{-1}). It clearly shows that as the shear rate increases, the viscosity of the abrasive media decreases, which implies that the abrasive media shows a shear thinning behaviour. At a lower shear rate of 0.01 s^{-1}, the viscosity of the abrasive media is 4.1×10^5 Pa.s and it is reduced to 69.7 Pa.s at a higher shear rate of 100 s^{-1}. The long molecules entangled starts disentangle as the shear rate increases and orientate along the flow direction. As the temperature increases to 35°C, 45°C, and 55°C, the media viscosity further reduces due to the weakened molecular force of attraction causing the easy flow of media shown in Figure 4.11 (b). The shear rate in the rotational rheometer with parallel plate configuration is determined by the angular velocity (ω), the radius of the plate (r), and the gap between two plates (h) (sample thickness) as described by the following equation (Mezger, 2006; Aho et al., 2015):

$$\text{Shear rate } (\dot{\gamma}) = \frac{\omega \, r}{h} \tag{4.14}$$

The viscosity of the abrasive media is given by Eq. (3.6)

$$\text{Viscosity } (\eta) = \frac{2 \, M}{\dot{\gamma}} \, \pi \, r^3 \left[\frac{3}{4} + \frac{1}{4} \frac{d(\ln M)}{d(\ln \dot{\gamma})} \right] \tag{4.15}$$

where M – torque.

(a) Temperature: 25 °C

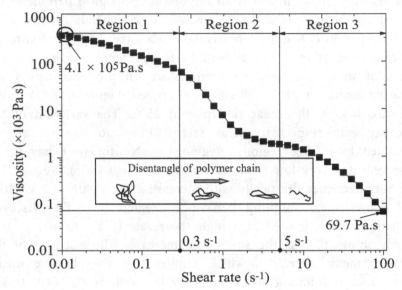

(b) Temperature: 35 °C, 45 °C and 55 °C

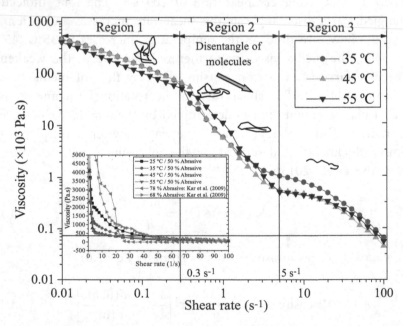

FIGURE 4.11 Viscosity vs. shear rate of the abrasive media at different temperatures.

From the trend (Figure 4.11), it is clear that as the shear rate and temperature increases, the viscosity of the media reduces; this shows the shear thinning or pseudoplastic behaviour of the media. This is due to a structural breakdown in the media as the shear rate and the temperature increase, and similar observations are reported in the literature (Cross, 1979; Kar et al., 2009b; Shit and Shah, 2013). As the temperature increases, the polymer long chains break into small segments, and also molecules gain energy and start moving apart. This will reduce the viscosity of the media, in turn, the radial force exerted on the abrasive particles decreases and results in reduced shear stress and affects the finishing performance (Sankar et al., 2011). The obtained results are also compared with available literature (shear rate 0 s^{-1} to 100 s^{-1}) and observed almost similar trend, and the results match with the media developed by Kar et al. (2009a) with 78% SiC abrasives and 68% of SiC abrasives.

4.6.5.2 Shear Stress vs. Shear Rate

Figure 4.12 shows the shear stress vs. shear rate plot of the abrasive media at a different temperature. The change in shear stress shows almost sinusoidal behaviour with respect to shear rate. From Figure 4.12 (a) at 25°C, three main regions are observed: Region 1 – linear behaviour is observed, as the shear rate increases from 0.01 s^{-1} to 0.3 s^{-1} and the shear stress also increases from 4,100 Pa to 21,500 Pa. Region 2 – shear stress decreases to 5,900 Pa with an increasing shear rate from 0.3 s^{-1} to 5 s^{-1}. Region 3 – shows the non-linear behaviour at a high shear rate from 5 s^{-1} to 100 s^{-1}, which exactly follows the transition points of the viscosity plot.

A similar trend is observed at temperatures 35°C, 45°C, and 55°C as shown in Figure 4.12 (b). Further, it is also observed that as the temperature increases, the shear stress of the abrasive media decreases. The rising part of the stress corresponds to the elastic deformation of the whole sample. When the stress reaches its maximum level, a fracture occurs that separates the sample into two different parts (upper block and lower block), and then the stress starts to fall. During the decreasing part of the stress, the two blocks lose the elastic strain that they have gained during the rising part of the stress. The upper block slips over the lower block through the fracture plane. The torsion bar brings the upper block back in the opposite direction of that of shearing. Meanwhile, the rotating lower plane carries the lower block along with it in

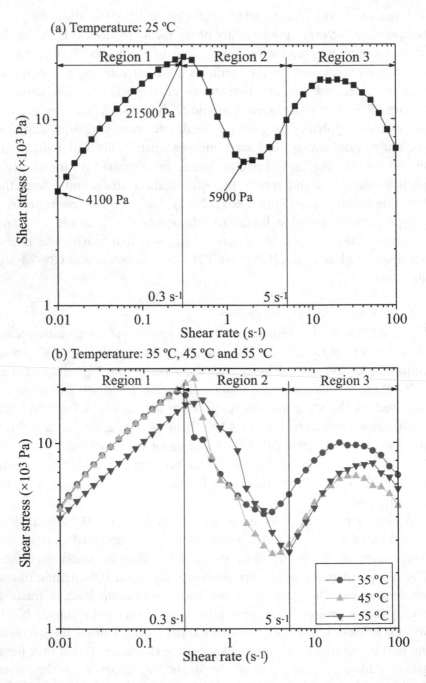

FIGURE 4.12 Shear stress vs. shear rate of the abrasive media at different temperature.

the direction of shearing. At the end of the falling part of the stress, the two blocks stick, they are again elastically strained, and the process starts again (Pignon et al., 1996). The shear stress (τ) for a parallel plate configuration with the rotational mode is determined by Eq. (4.16) (Mezger, 2006). Higher the media shear stresses, more are the material removal rate.

$$\text{Shear stress } (\tau) = \frac{2 M}{\pi r^3} \left[\frac{3}{4} + \frac{1}{4} \frac{d (\ln M)}{d (\ln \dot{\gamma})} \right] \tag{4.16}$$

4.6.5.3 Storage Modulus and Loss Modulus vs. Frequency

The storage modulus and the loss modulus give the details on the stress response of abrasive media in the oscillatory shear study. This study is also used to understand the microstructure of the abrasive media and to infer how strong the material is. Storage modulus (G') is a measure of the energy stored by the material during a cycle of deformation and represents the elastic behaviour of the material. Loss modulus (G'') is a measure of the energy dissipated or lost as heat during the shear cycle and represents the viscous behaviour of the material (Sankar et al., 2011). The terms G' and G'' can be expressed as sine and cosine function of the phase shift angle (δ).

$$\text{Storage modulus } (G') = \frac{2 h M_0 \cos\delta}{\pi r^4 \phi_0} \tag{4.17}$$

$$\text{Loss modulus } (G'') = \frac{2 h M_0 \sin\delta}{\pi r^4 \phi_0} \tag{4.18}$$

where M_0 is torque amplitude, ϕ_0 is the angular amplitude of oscillation and δ is the phase angle that represents the time-dependent behaviour of material that is a lag of reaction of the strain to the sinusoidal stress applied to the sample (Aho et al., 2015), and r is the radius of the parallel plate.

In general, storage modulus (G') and loss modulus (G'') are considered to distinguish the phases of materials considered for investigations. If $G'>G''$, it is a solid state, if $G'<G''$, it is a liquid state, and $G' = G''$, it is a gel state reported by Tseng et al. (2010). The effective material removal from the work surface depends on the state – almost solid when it interacts with the work surface. To understand the properties

of the developed media before using in the finishing process, storage modulus and loss modulus are examined.

Figure 4.13 shows the storage modulus (G') and loss modulus (G'') vs. frequency for various temperatures such as 25°C, 35°C, 45°C, and 55°C. The trend shows the storage modulus and the loss modulus of the abrasive media increases with an increase in frequency and decreases with an increase in temperature. Figure 4.13 (a) shows the results of the storage and loss modulus vs. frequency at temperature 25°C. The G' increases from 0.018 MPa to 0.77 MPa, and also, the G'' increases from 0.0187 MPa to 0.22 MPa as the frequency increases from 0.01 Hz to 100 Hz. Further, for different temperatures– 35°C, 45°C, and 55°C – the trend follows the same as 25°C as shown in Figure 4.13 (b). As the temperature increases, the G' and G'' decrease. The developed media behave like an elastic solid as because of $G'>G''$ at different temperatures with a varying frequency that is best suitable for the finishing process.

Storage modulus is solely responsible for the maximum material removal because it decides the radial force exerted by abrasive grain on the work surface. Also, more free space between the polymer chains will not give sufficient support to the abrasive particles while shearing the surface roughness peaks particles pushes back and rolls about its own axis. Studies conducted by Davies and Fletcher (1995), Kar et al. (2009a, 2009b), and Sankar et al. (2011) describe the improvement in the storage modulus and reduction in the free space between the polymer chains increases the efficiency of the media by providing the better shear strength characteristics. Low storage modulus reduces the shear strength, and high storage modulus reduces the abrasive media flow-ability. So, it is better to maintain the intermediate storage modulus that can increase the abrasive media performance during the finishing process (Sankar et al., 2011).

At lower frequency, the storage modulus is lesser than the loss modulus; it means viscous property of the media dominates the elastic property. As the frequency increases, the storage modulus increases; it shows the abrasive media has the capacity to store more energy, and it crosses loss modulus at a point called cross-over point. The cross-over point of the modulus is related to the molecular architecture of the polymer and molecular weight distribution (MWD). Polymers with a high molecular weight (M_w) have the crossover (shift from viscous to elastic dominated behaviour) at a lower frequency compared to a low molecular weight polymer (Aho et al., 2015).

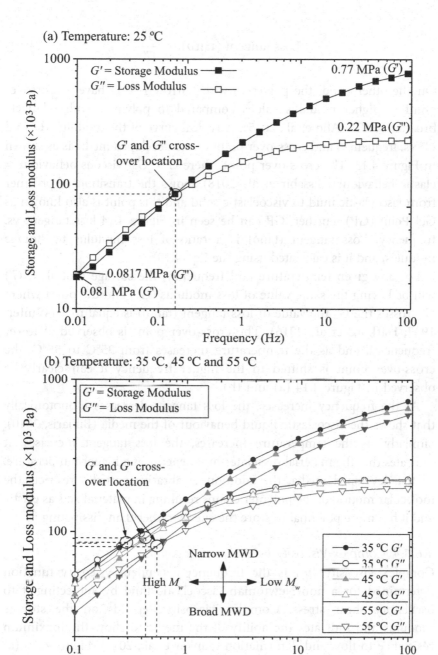

FIGURE 4.13 Storage and loss modulus vs. frequency of the abrasive media at different temperatures.

$$\text{Loss tangent } (\tan \delta) = \frac{G''}{G'} \qquad (4.19)$$

On the other hand, the polymers with narrow MWD have a crossover point at higher modulus values compared to polymers with relatively broader MWD (Aho et al., 2015). A typical curve of the modulus (G' and G'' vs. frequency) and cross-over point of the developed media is as shown in Figure 4.13. The cross-over point where $G' = G''$ (viscous behaviour = elastic behaviour) (Essabir et al., 2018) shows the transition of polymer from viscoelastic fluid to viscoelastic solid and this point is also known as Gel Point (GP). Further, GP can be seen in Figure 4.14 loss tangent vs. frequency. Loss tangent ($\tan\delta$) is a ratio of loss modulus to storage modulus, and it is calculated using the Eq. (4.19).

For any given temperature and frequency, the storage modulus (G') will be having the same value of loss modulus (G'') and the point where G' crosses the G'' the value of loss tangent ($\tan \delta$) is equal to 1 (Winter, 1987; Harkous et al., 2016). The cross-over point is observed at lower frequencies, and as the temperature increases from 35°C to 55°C, the cross-over point is shifted to the higher frequency it can clearly be observed in Figure 4.14 (a) and (b).

As the frequency increases, the loss tangent decreases monotonically that shows the viscoelastic liquid behaviour of the media (Bikiaris, 2010). Similarly, as the temperature increases, the loss tangent increases; it indicates that the material dissipates more energy at a higher temperature. In the developed media, the addition of abrasive particles restricts the molecular motion of the polymer chains resulting in material acts as elastic and it has more potential to store the energy rather than dissipating it.

4.6.5.4 Complex Viscosity vs. Frequency

Complex viscosity (η^*) is the frequency-dependent viscosity function determined for a non-Newtonian viscoelastic fluid by subjecting it to oscillatory shear stress. Complex viscosity depends on the storage modulus and indicates the ability of the media to show the maximum resistance to flow and deformation (Sankar et al., 2011). Figure 4.15 (a) shows the complex viscosity vs. frequency plot of media at 25°C. The media shows non-Newtonian behaviour (shear thinning) because the decrease in complex viscosity is observed with an increase in frequency. At the lowest temperature of 25°C, the complex viscosity of media

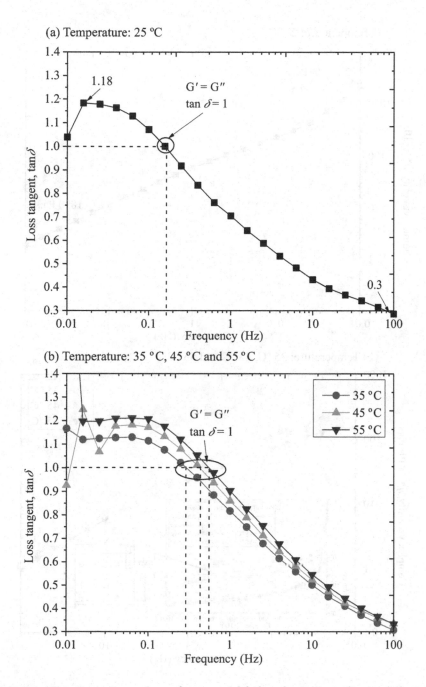

FIGURE 4.14 Loss tangent vs. frequency of the abrasive media at different temperatures.

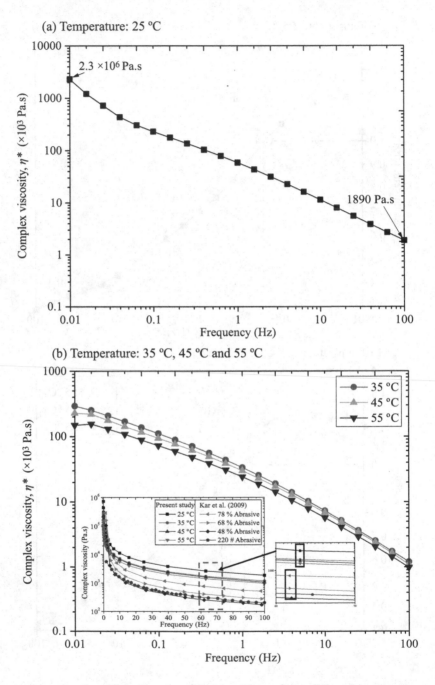

FIGURE 4.15   Complex viscosity of the abrasive media at different temperatures.

is 2.3×10^6 Pa.s, and it is reduced to 1,890 Pa.s as the frequency increases from 0.01 Hz to 100 Hz. A similar trend is observed at 35°C, 45°C, and 55°C shown in Figure 4.15 (b). At the highest temperature of 55°C, the complex viscosity of media is reduced from 0.149×10^6 Pa.s to 983 Pa.s. This variation is due to the conversion of elastic energy of the media to the viscous energy. The elastic modulus dominates the media behaviour, and the study shows the media should be sheared at a very high shear rate to obtain the low viscosity for easy flow (Kar et al., 2009a). As the frequency increases, this breaks the polymer chain. Similarly, as the temperature increases, polymer starts breaking (chain scission) and weakens the bonding between the polymer and abrasive. This alter in polymer chain allows the media to flow easily and, in turn, affects the force required to remove the material from the workpiece and reduces the finishing process efficiency.

The normal working temperature of the finishing process is less than 70°C. This temperature rise is due to friction between the work surface and the resistance of the particles to applied pressure and also continues usage of the media. Under this temperature range, a significant effect of temperature on the rheological properties of media is observed, and also it is a significant factor to be considered in finishing process study. The results are compared with the available literature with different percentages of abrasives and 220 mesh size of particles developed and characterized by Kar et al. (2009b) under room temperature (25°C), and it reveals developed media shows better properties.

The complex viscosity (η^*) is defined as:

$$\text{Complex viscosity } (\eta^*) = \sqrt{(\eta')^2 + (\eta'')^2} \qquad (4.20)$$

where η' is the dynamic viscosity, which is the ratio of the stress in-phase with the rate of strain to the amplitude of the rate of strain; η'' is the out-of-phase viscosity, which is the ratio of the stress 90° out of phase with the rate of strain to the amplitude of the rate of strain in the forced oscillation as given by Eq. (4.21) and Eq. (4.22) measured at angular velocity, $\omega = 2\pi f$.

$$\text{Dynamic viscosity } (\eta') = \frac{G''}{\omega} \text{ (viscous portion of complex viscosity)}$$

$$(4.21)$$

$$\text{Out-of-phase dynamic viscosity } (\eta'') = \frac{G'}{\omega} \tag{4.22}$$
$$\text{(elastic portion of complex viscosity)}$$

After the thorough investigations on thermal and rheological character-istics of the developed 220 mesh size abrasive media, it is subjected to the finishing of various different materials having different hardness number and also the complex internal features.

4.7 SUMMARY

The development of the UAFF process is carried out in two stages. Initially, UAFF experimental setup is developed in which the abrasive media flows in one direction, and after one cycle, the abrasive media is fed back to the machine manually. To avoid the manual filling of abrasive media, stage 2 experimental setup is developed in which the closed loop concept is used to achieve the automatic recirculation of the abrasive media. Different work sample geometries and fixtures are developed to carry out the experimentation that is presented in this chapter. Abrasive media used in the UAFF process acts as a flexible tool that can reach any complex internal and external surfaces and cavities available on the work surfaces. The development of abrasive media for this application is challenging, and proper selection and blending of the different constituents and compositions of the abrasive media requires a lot of attention. The silicone rubber, silicone oil, and silicon carbide particles are selected as constituents of the abrasive media in the present work. Two roll mill machine is used to blend all these constituents and abrasive media with varying abrasive particle mesh size. Further, the developed media is subjected to different studies such as morphology, thermal stability, and rheological properties to understand media behaviour. The properties are also compared with literature and observed a similar trend. These abrasive media are used to finish the different materials, and features and results are presented and discussed in Chapter 5.

CHAPTER 5

Effect of UAFF Process Parameters on Wettability and Bacterial Adhesion

5.1 INTRODUCTION

The experimentations have been carried out at various levels to investigate the performance of the developed unidirectional abrasive flow finishing (UAFF) experimental setup. The effect of process parameters such as a number of cycles, pressure and abrasive media size on average surface roughness (R_a), surface morphology and geometry of the components is studied. The effect of surface roughness on wettability and bacterial adhesion of the machined work surfaces is examined, and obtained results are presented in this Chapter.

5.2 MACHINABILITY STUDY

The main objective of this work is to study the effect of process parameters on finishing four different engineering materials such as aluminium, brass, copper and mild steel selected based on their hardness properties listed in Table 5.1. Material selection, workpiece details and fixture details are explained in Chapter 4. Experiments are

TABLE 5.1   Experimental details and results obtained during the machinability study

Parameters	Particulars						
Rectangular workpiece (l × b × t)	Al (99 HB), Brass (140 HB), Cu (85 HB), MS (202 HB) 40 mm × 5 mm × 6 mm						
Abrasive media	Silicone rubber (38%) + Silicone oil (12%) + SiC (50%)						
Pressure and stroke length	50 bar and 400 mm						
Achieved average surface roughness, R_a(µm)							
No. of cycles in⟶	0	3	6	9	12	% impro.	Rank
Aluminium (Al)	3.74	3.13	2.29	1.86	1.51	59.62%	2
Brass	1.56	1.23	0.91	0.78	0.70	56.12%	3
Copper (Cu)	1.73	1.58	1.09	0.80	0.50	70.52%	1
Mild Steel (MS)	2.06	1.62	1.46	1.29	1.10	46.53%	4

conducted to remove the irregularities present on these milled surfaces and studied the effect of a number of cycles in achieving the different roughness parameters. Table 5.1 shows the experimental details and obtained results during the machinability study. The change in surface roughness is measured at varying number of cycles. It is observed that as the number of cycles increases, the R_a value reduces. Surface roughness value varies nonlinearly with an increase in a number of cycles irrespective of the type of material. Initially, for the first few numbers of cycles, the drastic changes in roughness value are observed. Still, later, progressive reduction in roughness values is noted with an increasing number of cycles. Because the machined surfaces will be having more sharp edges, when the abrasive media is exposed to these peaks, the peaks get machined due to abrasion. After the first few cycles, the density of sharp peaks reduces, and surface attains a critical surface roughness value.

From the experiments, it is observed that the percentage improvement in surface roughness for different engineering materials such as aluminium, brass, copper and mild steel is 59.62%, 56.12%, 70.52% and 46.53%, respectively. The obtained results show that the maximum percentage improvement in R_a of 70.52% is observed for copper because of its lower hardness of 85 HB among all the other selected materials. Minimum percentage improvement in R_a of 46.53% is observed for mild steel, which has a higher hardness of 202 HB. Hence, the abrasive flow finishing process is very useful in finishing of soft materials as well as hard materials.

5.3 EFFECT OF SURFACE ROUGHNESS ON WETTABILITY STUDY

Despite the adoption of advanced technology in manufacturing implants and surgical and medical management procedures, there are still a large number of implants that are subject to failure. The American joint replacement registry annual report shows that the major causes for implants failure are a fracture, prosthetic dislocation, loosening, excessive wear rate at mating surfaces, and its associated debris and pre-surgical contamination/infection (i.e., bacterial adhesion). Surfaces roughness is one of the factors which influence bacterial adhesion on implants and acts as favourable sites for colonization and biofilm formation, which leads to prosthetic implant infections (PIIs). Most of the implants have a complex and partly freeform surface, which is difficult to access by the conventional tools in achieving the uniform surface quality.

Most popular metallic biomaterials in use today are stainless steel, cobalt-chromium-molybdenum alloys, pure titanium and titanium alloys-Ti-6Al-4V. Stainless steel and cobalt-chromium-molybdenum alloys have higher Young's modulus around 200 GPa, while titanium alloys – Ti-6Al-4V – have Young's modulus around 110 GPa which is almost close to the bone. Also, the density of Ti-6Al-4V is about 4,500 kg/m^3 which is less compared to SS316 (7,900 kg/m^3) and Co-Cr-Mo alloys (8,300 kg/m^3), and made Ti-6Al-4V material as the best option for manufacturing orthopaedic implants. In the present work, SS316L and Ti-6Al-4V workpiece materials are selected to show the ability of unidirectional abrasive flow finishing (UAFF) process in finishing of these biomaterials. The objective of this experiment is to finish the biomaterials and to study the effect of the UAFF process on the surface roughness, surface morphology, bearing area curve (BAC) and wettability of the biomaterial work surfaces. For experimentation purposes, the rectangular shape samples are prepared with the dimensions of 20 mm × 10 mm × 6 mmas shown in Figure 5.1(a), and these are cut using a wire electric discharge machine. Prior to the finishing process, the samples are polished using metallographic SiC papers having grit size 220 (8-inch diameter) placed in the polishing machine for 120 s time duration, with a speed of 150 rpm. Viscoelastic polymer-based flexible abrasive media having the concentration of SiC of mesh size 320 (50%) + silicone rubber (38%) + silicon oil (12%) were used to finish these surfaces. The selected process parameters are as follows: a number of cycles = 3, 6 and 9, and

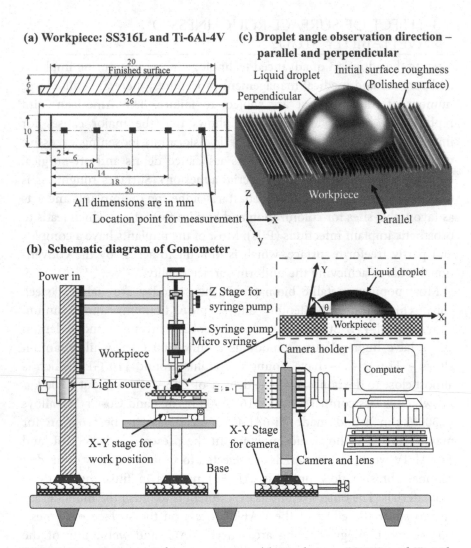

FIGURE 5.1 Contact angle measurement. (a) Workpiece: SS316L and Ti-6Al-4V, (b) Schematic diagram of goniometer and (c) Droplet angle observation direction –parallel and perpendicular.

pressure = 40 bar, 50 bar and 60 bar. The obtained surface roughness and BAC are measured using an optical profilometer, and the morphology of the surface is studied using a scanning electron microscope (SEM).

Further, the wettability of the finished surface is studied by measuring the contact angle (θ) for three different liquids – water, formamide and diiodomethane by sessile drop technique using the contact angle Goniometer instrument. The measured contact angles are used to ascertain the surface free energy components using a van-Oss-Chaudhury-Good equation. The significant difference is observed on surface roughness, contact angle and surface energy of the machined surfaces at different finishing cycles. Also, the various tendencies of the droplet contact angle and surface energy have been observed along the finishing direction parallel and perpendicular, and it ascertained the firm conclusion that surface roughness and surface morphology play a significant role in wetting characteristics. Further, the response surface methodology (RSM) model has been used to optimize the input parameters such as a number of cycles and pressure to obtain desired output responses, namely, R_a and material removed (MR). Interactive effects of a number of cycles and pressure on the R_a and MR are discussed in this section.

5.3.1 Characterization of Surface Roughness and Wettability

Optical profilometer (Wyko NT1100, Veeco Instruments, USA) is used to measure R_a and BAC with full resolution and threshold of 0.35%, scan length of 50 µm, a back scan of 10 µm and scanning area of 612.6 µm × 466.1 µm. Utmost care has been taken while measuring the parameters. The samples are cleaned with ethanol, and five measurements on each sample are taken before and after finishing. The variation in R_a is represented with an error bar based on the standard deviation.

Goniometer (GBX-Digidrop MCAT) device is used for contact angle measurement. A schematic diagram of the setup is shown in Figure 5.1 (b). Initially, the surfaces are cleaned with distilled water and ultrasonicated for 5 min with ethanol using ultrasonicator and dried at room temperature before taking the measurements. The droplet volume of liquid is made within the range where the contact angle did not change with the change in droplet volume. Deionized water, formamide and diiodomethane liquid droplets of 2 µl (+ 0.2 µl) volume are placed on each sample using a microsyringe, and the droplet contact angle is measured at the room temperature of 25°C and the humidity of 71%. The readings are taken in parallel and perpendicular directions of the abrasive media flow on the work surfaces, as shown in Figure 5.1 (c). The delay time of measurement in the experiment is set to be 20 s, and the images are captured up to 60 s to measure any change in droplet angle and volume with respect to time. Measurements are repeated five times on each work sample, as shown in Figure 5.1 (a), and the variation is found within the range of ± 1° to 5°.

5.3.2 Surface Roughness and Surface Morphology of Finished Surfaces

Figure 5.2 shows the effect of pressure and the number of cycles on the surface roughness of the machined materials – SS316L and Ti-6Al-4V. From the experiments, it is evident that, as the number of cycles and pressure increases, the surface roughness decreases by removing the irregularities present over the work surfaces. Figure 5.2 (a) shows the effect of pressure and number of cycles on the SS316L material. Initially,

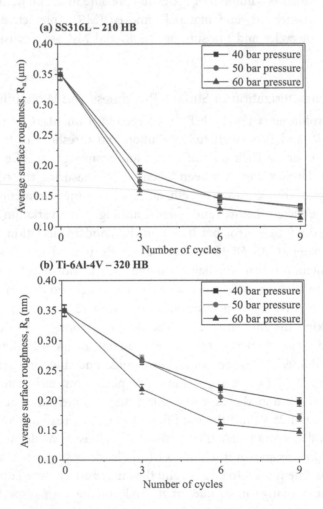

FIGURE 5.2   Effect of pressure and number of cycles on average surface roughness. (a) SS316L – 210 HB and (b) Ti-6Al-4V – 320 HB.

the surface roughness value of the SS316L is 0.350 µm. After finishing with 3 cycles at 40 bar pressure, surface roughness reduced to 0.193 µm. Further, for the next 6 cycles and 9 cycles, surface roughness reduced to 0.145 µm and 0.133 µm, respectively. It is observed that during the initial cycles, that is, the first 3 cycles, the change in surface roughness is more, after 6 cycles, the change in surface roughness is less. Similar results are observed in the case of 50 bar pressure and 60 bar pressure. Finally, the minimum surface roughness 0.114 µm is achieved on the surface finished with 60 bar pressure at 9 cycles for SS316L. With respect to pressure, as the pressure increases, the change in surface roughness is more during the initial cycles. As the number of cycles progresses with an increase in the pressure, even though the surface roughness is reduced, the difference of surface roughness at two pressure is less. Similarly, Figure 5.2 (b) shows the effect of pressure and number of cycles on the Ti-6Al-4V material. Before finishing, the surface roughness is found to be 0.350 µm, later as the number of cycles increases, and the surface roughness reduces to 0.267 µm at 3 cycles, 0.220 µm at 6 cycles and 0.197 µm at 9 cycles for 40 bar pressure. The minimum surface roughness of 0.147 µm is achieved at 60 bar pressure and 9 cycles.

During the initial cycles of operation, irrespective of pressure, the reduction in surface roughness is much higher. As the number of cycles increased, reduction in surface roughness is less. The possible reasons for these effects are, initially, machined surfaces will be having more sharp peaks and valleys when the abrasive media comes in contact with these peaks, peaks get machined due to abrasion, and hence peaks reduce drastically as media move over the work surfaces. After a few cycles, the density of sharp peaks of the work surfaces reduces, and the surface attains a critical surface roughness value. So, once the surface reaches the critical surface roughness value, the finishing is not possible drastically further, and the surface becomes somewhat flatter than before, as also seen from the published literature (Jain and Adsul, 2000; Walia et al., 2006a). Comparing the results, initially, both the materials – SS316L and Ti-6Al-4V– had almost the same R_a but after finishing, SS316L achieved better surface finish as compared to Ti-6Al-4V within the conducted experimental range. This may be because the hardness of SS316L (210 BHN) is less than the hardness of the Ti-6Al-4V (320 BHN) material. Hence, the material hardness is also one of the important parameters which affect the finishing cycles, and it is also confirmed during the machinability study.

The obtained surface roughness value gives only the information about variation in the height of the irregularities, but it does not give any information about sizes, slopes and shapes of surface irregularities. Lay patterns with linear grooves and crisscross grooves have an influence on the friction between surfaces. A possible way to describe these types of surfaces is by using BAC. BAC also called as Abbott–Firestone curve is used for the analysis of load carrying surfaces (Kanthababu et al., 2009; Laheurte et al., 2012; Bhushan, 2013). It is used to show the quality of surfaces, which gives information about the size and proportions of the peaks and valleys present on the work surfaces (Rohm et al., 2016). In the present work, an attempt has been made to show the BAC of SS316L and Ti-6Al-4V graphically before and after finishing. Figure 5.3 shows the variation of the BAC at different pressure and number of cycles. From the plot, one can analyse, as the number of cycles and pressure increase, the bearing area under the curve (which gives the information on the density of peaks and valleys) decreases irrespective of the materials.

The surface morphology of the machined surfaces is examined using a high-resolution scanning electron microscope (HRSEM) – Inspect F, as shown in Figure 5.4. Figure 5.4 (a) shows the HRSEM images of the SS316L before finishing and at different stages of finishing cycles at 60 bar pressure. It can be observed that initially deep machining marks, burn spots and pits are observed on the machined surfaces. As the number of cycles increases to 3, 6 and 9, the machined marks and surface damages are removed, and new lay patterns are formed with the better surface finish. Also, it has been observed that before finishing, the lay patterns are parallel to the polished direction, and after finishing, the lay patterns changed to a direction parallel to the abrasive media flow direction. Similarly, Figure 5.4 (b) shows the SEM images of the Ti-6Al-4V before finishing and at a different number of cycles at 60 bar pressure.

Figure 5.5 shows the close-up view of machined work surfaces before and after finishing. The surface finish obtained on the work surfaces shows the reflection of the letters; that is, the mirror finish is achieved on the machined surfaces. Further, the analysis is extended to understand how the surface roughness and surface morphology of the finished surfaces affect the wettability characteristics. The obtained results obtained are discussed in the next section.

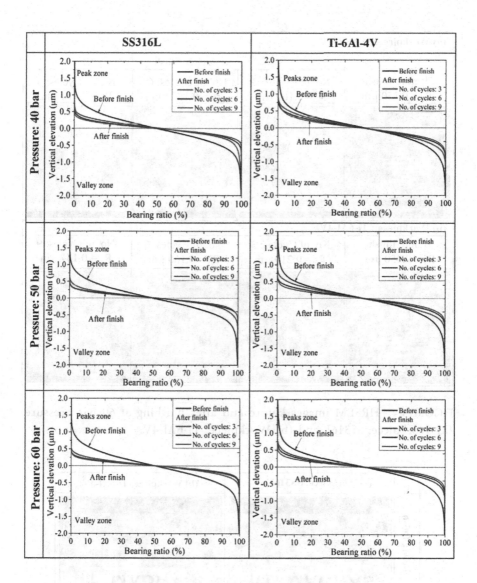

FIGURE 5.3   Variation of bearing area curve at different pressure and number of cycles.

5.3.3 Wettability Study – Contact Angle and Surface Free Energy

The wettability of SS316L and Ti-6Al-4V of the machined work surfaces is investigated through contact angle measurement for three different liquids – water, formamide and diiodomethane. Figure 5.6 shows the

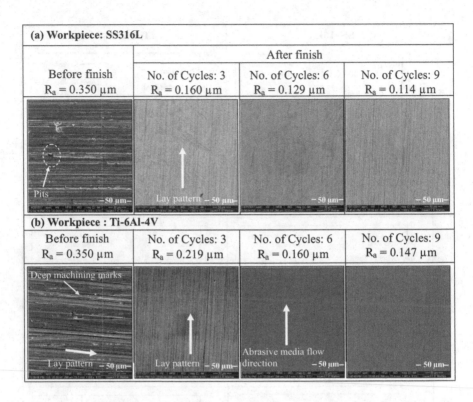

(a) Workpiece: SS316L			
	After finish		
Before finish $R_a = 0.350\ \mu m$	No. of Cycles: 3 $R_a = 0.160\ \mu m$	No. of Cycles: 6 $R_a = 0.129\ \mu m$	No. of Cycles: 9 $R_a = 0.114\ \mu m$
Pits	Lay pattern		
(b) Workpiece : Ti-6Al-4V			
Before finish $R_a = 0.350\ \mu m$	No. of Cycles: 3 $R_a = 0.219\ \mu m$	No. of Cycles: 6 $R_a = 0.160\ \mu m$	No. of Cycles: 9 $R_a = 0.147\ \mu m$
Deep machining marks Lay pattern	Lay pattern	Abrasive media flow direction	

FIGURE 5.4 HRSEM images before and after finishing at 60 bar pressure. (a) Workpiece: SS316L and (b) Workpiece: Ti-6Al-4V.

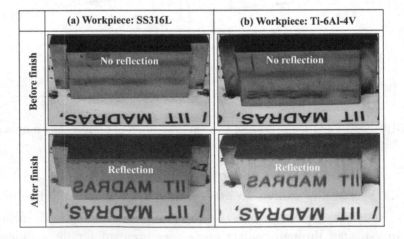

FIGURE 5.5 Close-up view of the machined surfaces before and after finishing. (a) Workpiece: SS316L and (b) Workpiece: Ti-6Al-4V.

effect of finishing on the contact angle for the SS316L workpiece. The droplet contact angle of water with the surface before finishing (onboard) is 103.6° along X-axis ($\theta^{\parallel}$) and 92.2° along Y-axis ($\theta^{\perp}$).

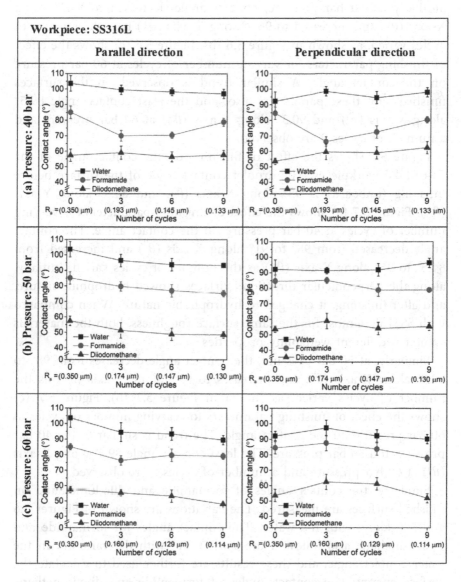

FIGURE 5.6 Effect of finishing on contact angle – SS316L. (a) Pressure: 40 bar, (b) Pressure: 50 bar and (c) Pressure: 60 bar.

Figure 5.6 (a) shows the effect of finishing parameters for varying number of cycles at 40 bar pressure on contact angle (θ). As the number of cycles increases from 3 to 9, contact angle decreases from 103.6° to 97° along X-axis ($\theta^{||}$) and increases from 92.2° to 98° along Y-axis ($\theta^{\perp}$). Similarly, at 50 bar pressure, contact angle decreased to 93.2° along X-axis ($\theta^{||}$) and increased to 96.3°along Y-axis ($\theta^{\perp}$) it as the number of cycles varied, as shown in Figure 5.6 (b). Figure 5.6 (c) shows the effect of finishing parameters for varying number of cycles at 60 bar pressure on the contact angle. A similar trend is observed on the surfaces finished with these parameters also, and the least contact angle 89.5° along X-axis ($\theta^{||}$) and 90.3° along Y-axis ($\theta^{\perp}$) at 60 bar pressure and a number of cycles 9 are observed.

Figure 5.7 shows the effect of finishing on the contact angle of the Ti-6Al-4V workpieces. The droplet contact angle of the surface before finishing (onboard) is 98° along X-axis ($\theta^{||}$) and 84.5° along Y-axis ($\theta^{\perp}$). Figure 5.7 (a) shows the effect of finishing parameters for varying number of cycles at 40 bar pressure on the contact angle. The contact angle decreased from 98° to 91° along X-axis ($\theta^{||}$) and increased from 84.5° to 95° along Y-axis ($\theta^{\perp}$) as the number of cycles varied. Initially, along the perpendicular direction, surfaces showed hydrophilic nature, and after finishing, it changed to hydrophobic nature. When compared with SS316L, almost for the same surface roughness, both the materials exhibited different wettability properties.

Similarly, at 50 bar pressure, the contact angle decreases from 98° to 90.9° along X-axis ($\theta^{||}$) and increases to 93.3°along Y-axis ($\theta^{\perp}$) as the number of cycles varied, as shown in Figure 5.7 (b). Figure 5.7 (c) shows the effect of finishing parameters for varying number of cycles at 60 bar pressure on the contact angle. The trend is similar to the 40 bar pressure and 50 bar pressure. The least contact angle 89.3° along X-axis ($\theta^{||}$) at 60 bar pressure and a number of cycles 9 are observed.

Similarly, the contact angles of formamide and diiodomethane on finished surfaces are measured. The variations are shown in Figures 5.6 and 5.7, along with the water. The contact angles for formamide and diiodomethane also followed the same trend with the variation in the water contact angle, and these results are further used to calculate the surface energy. The contact angles of formamide and diiodomethane with a solid surface are less compared to water because the surface tension of water is more compared to these liquids. From the contact angle results, it could be stated that the spreading phenomenon is

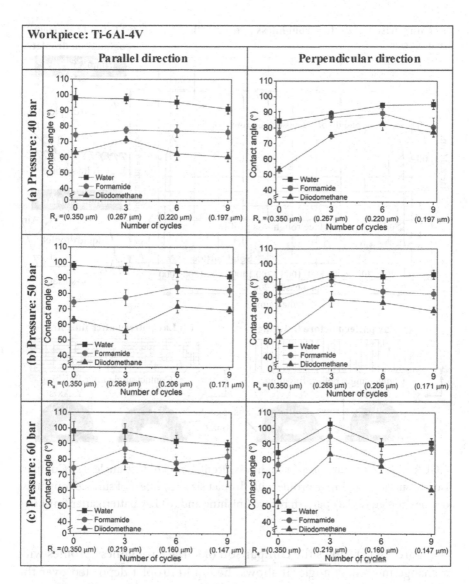

FIGURE 5.7 Effect of finishing on contact angle – Ti-6Al-4V. (a) Pressure: 40 bar, (b) Pressure: 50 bar and (c) Pressure: 60 bar.

different along X-axis and Y-axis in the case of both the materials and this affects the change in contact angle. The possible reason for this phenomenon is surface roughness and lay patterns direction on the work surface and is explained below.

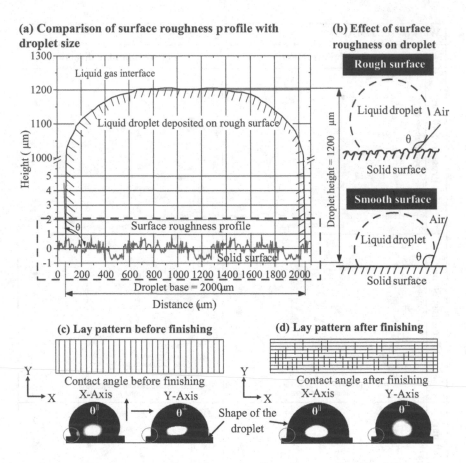

FIGURE 5.8 Comparison of roughness profile and contact angle. (a) Comparison of surface roughness profile with droplet size, (b) Effect of surface roughness on the droplet, (c) Lay pattern before finishing and (d) Lay pattern after finishing.

Figure 5.8 (a) shows the comparison of surface roughness profiles with a change in contact angle. It shows the liquid droplet deposited over the work surface covering the several small peaks and valleys, and the same thing is also seen from the literature (Kubiak et al., 2011). Figure 5.8 (b) illustrates the effect of surface roughness on a droplet of liquid, which is spreading on the machined surfaces. The reported results of the contact angle can be correlated to surface roughness and surface morphology. It is observed from the results that the contact angle is strongly influenced by the surface lay direction of the finished surfaces because the measured contact angles on the work surface along two directions are different.

This shows the anisotropy nature of the machined surface as detailed in the study conducted (Fischer et al., 2014; Li et al., 2014; Liang et al., 2015). Before finishing, the contact angle along the X-axis (parallel direction, $\underline{\theta}^{||}$) is larger and that along the Y-axis (perpendicular direction, $\underline{\theta}^{\perp}$) is smaller. So, it is obvious that the grooves present on the surface do not allow the droplet to spread in the perpendicular direction to the groove direction because it creates an energy barrier to the movement of the water droplet. As a result, the pinning of a droplet on the work surface is observed (Liang et al., 2015; May et al., 2015; Cheng et al., 2016). The droplet spread along the direction parallel to the groove direction because they get the free flow path indicating the smaller contact angle. Also, it is seen from the literature that surface anisotropy has a strong influence on the directional spreading and wettability (Cunha et al., 2013; Fischer et al., 2014; Li et al., 2014; Liang et al., 2015). Figure 5.8 (c) shows the schematic diagram of the surface lay pattern before finishing. Later, after finishing with the present process, the contact angle decreases in X-axis because initially, the peaks' heights are more; as the finishing cycles increase, the height of these peaks reduces, and it results in the decrease in surface roughness and contact angle. Also, the groove's directions change in the finishing direction.

It is observed that on the Y-axis (perpendicular direction), this phenomenon is completely opposite because before finishing, the grooves are perpendicular to the finishing direction. After finishing, the direction of the grooves changes along the direction parallel to the finishing direction. At initial stages, the contact angle increases, and later, it decreases on some surfaces due to irregular lay patterns, as illustrated in Figure 5.8 (d). It is believed that during the first few cycles, the abrasives may create microgrooves like a plus (+) mark at some places. Due to this irregular change in lay pattern, contact angle may increase and may later decrease, and it has a great influence on the surface energy.

Table 5.2 shows the consolidated results of the change in contact angle with different finishing process parameters. A sharp decrease in the $\Delta\theta$ is observed when surface roughness decreases, and this is observed in both the materials. It shows the anisotropic wetting of the surface, so the contact angle decreases with a decrease in the surface roughness. The $\Delta\theta$ of 11.4° on SS316L and 13.5° on Ti-6Al-4V is observed on the workpiece surface before finishing. Later, as the finishing process progresses, the $\Delta\theta$ starts reducing and minimum $\Delta\theta$ is found on the workpiece surface with lower roughness values.

TABLE 5.2   Change in contact angle with different finishing process parameters

		SS316L		Ti-6Al-4V	
		R_a(μm)	$\Delta\theta$(deg)	R_a(μm)	$\Delta\theta$(deg)
As-received		0.350	11.4	0.350	13.5
	3 cycles	0.193	1	0.267	8.5
40 bar pressure	6 cycles	0.145	3.8	0.220	0.8
	9 cycles	0.133	1	0.197	4
	3 cycles	0.174	5.1	0.268	3.5
50 bar pressure	6 cycles	0.147	1.9	0.206	2.8
	9 cycles	0.130	3.1	0.171	2.4
	3 cycles	0.160	2.8	0.219	4.9
60 bar pressure	6 cycles	0.129	1.1	0.160	2
	9 cycles	0.114	0.8	0.147	1.3

R_a _ Average surface roughness; $\Delta\theta$ – change in contact angle

The result shows that the surface roughness along with surface lay patterns play a significant role in the wetting. Almost all the finished surfaces show the isotropic properties, that is, less $\Delta\theta$ compared to initial workpiece surfaces due to the reduced height of grooves, which in turn reduces the surface area and energy. For the biomedical orthopaedic applications, this property is significant to avoid the adhesion of bacteria along one direction, which is later challenging to remove. These surfaces also find the wide applications in microfluidic, heat transfer and lubrication of parts machined with high accuracy and high precision.

Figure 5.9 (a) shows the surface lay patterns, surface roughness profile and contact angle images before finishing (onboard) and for different machining cycles for SS316L. Initially, the surface is having the peaks in a range of 1.170 μm, and the lay pattern is oriented along the polished direction. It can also be seen from the surface roughness profile that the large irregular sizes of peaks and valleys are present over the surfaces (onboard). The contact angle images for different liquids such as water, formamide and diiodomethane clearly indicate the directional wetting along with parallel and perpendicular directions. Figure 5.9 (b–d) shows the surface lay patterns, surface roughness profile and contact angle images after finishing with varying number of cycles with 60 bar pressure. The surface roughness profiles indicate the

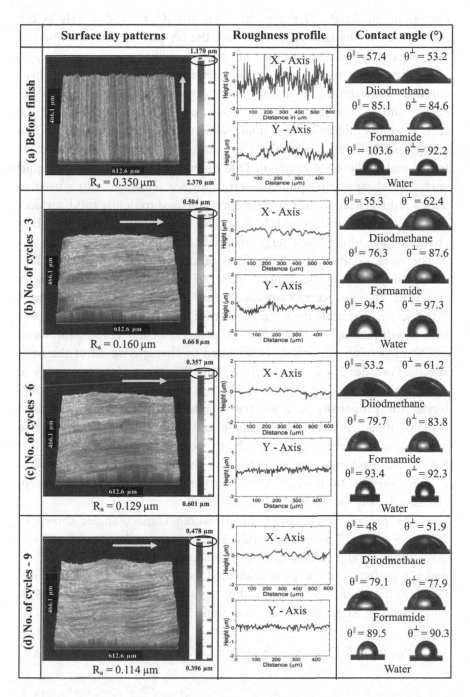

FIGURE 5.9 Surface lay patterns, surface roughness profile and contact angle images before and after finishing – SS316L material (at 60 bar pressure). (a) Before the finish, (b) No. of cycles – 3, (c) No. of cycles – 6, (d) No. of cycles – 9.

change in the height of the peaks along the X-axis and Y-axis as the finishing cycle changes. After finishing, the lay patterns are changing in the direction parallel to the flow of abrasive media.

Further, the effect of process parameters on surface energy is measured to understand the surface roughness and the lay pattern. Surface energy and surface tension result from non-symmetric bonding of the surface atoms/molecules in contact with vapour and are measured as energy per unit area, and these are measurements of intermolecular forces that make up a material. Due to intermolecular forces, a liquid surface is always being pulled inward. Surface tension is often used to define fluid surfaces, while surface energy is used to define solid surfaces (Mattox, 2010). Surfaces with high surface energy will try to lower their energy by absorbing low energy materials.

The surface energy of solid surfaces cannot be measured directly. It is calculated based on the contact angle of various liquid with surfaces measured. The most commonly used theory to calculate is van-Oss theory. It focuses on separating the surface energy of a solid into three components, such as a dispersive component, an acid component and a base component.

The dispersive component is intended to characterize all of the non-specific (van der Waals type) interactions that the surface is capable of making with a wetting liquid. The acid component, in theory, characterizes the propensity of a surface to interact by specific interactions (dipole-dipole, induced dipole-dipole and hydrogen bonding type) to wetting liquids, which have the ability to donate electron density (act basic). The base component, in theory, characterizes the propensity of a solid surface to interact with wetting liquids, which can accept electron density (act acidic) (Rulison, 1999).

In the present work, using the measured contact angles shown in Figures 5.6 and 5.7 and known values of surface tension components of three liquids water, formamide and diiodomethane (Table 5.3: γ^{LW}, γ^{+} and γ^{-}), surface free energy is calculated using the van-Oss-Chaudhury-Good equation as detailed in the literature (van-Oss et al., 1988; Ghanbari and Attar, 2014; Law and Zhao, 2016; Yan et al., 2017).

$$\left[(1 + \cos\theta_W)\, \gamma_W^{TOT}\right] = 2\left[\left(\gamma_S^{LW}\, \gamma_W^{LW}\right)^{0.5} + \left(\gamma_S^{+}\, \gamma_W^{-}\right)^{0.5} + \left(\gamma_S^{-}\, \gamma_W^{+}\right)^{0.5}\right] \quad (5.1)$$

$$\left[(1 + \cos\theta_F)\, \gamma_F^{TOT}\right] = 2\left[\left(\gamma_S^{LW}\, \gamma_F^{LW}\right)^{0.5} + \left(\gamma_S^{+}\, \gamma_F^{-}\right)^{0.5} + \left(\gamma_S^{-}\, \gamma_F^{+}\right)^{0.5}\right] \quad (5.2)$$

TABLE 5.3   Surface tension components of the liquids used for surface energy measurement (Wu, 1982; Rulison, 1999; Ghanbari and Attar, 2014)

Liquids	Total surface tension, γ^{TOT}(mN/m)	Surface tension components		
		γ^{LW}(mN/m)	γ^{+} (mN/m)	γ^{-}(mN/m)
Water (W)	72.8	21.8	25.5	25.5
Formamide (F)	58	39	2.28	39.6
Diidomethane (D)	50.8	50.8	0.1	0

γ^{TOT}– total surface tension of the liquid; γ^{LW}– Lifshitz–van der Waals component also known as dispersive component; γ^{+} ‾acid component, γ^{--} base component; S – solid; W – water; F – formamide; D – diiodomethane; θ– contact angle

$$[(1 + \cos \theta_D) \, \gamma_D^{TOT}] \;=\; 2\left[(\gamma_S^{LW} \, \gamma_D^{LW})^{0.5} + (\gamma_S^{+} \, \gamma_D^{-})^{0.5} + (\gamma_S^{-} \, \gamma_D^{+})^{0.5}\right] \quad (5.3)$$

For each workpiece, the contact angle values are substituted in the equation with respect to different liquids to form three non-linear equations. By solving these equations, three unknown solid surface energy components ($\gamma_S^{\,LW}$, γ_S^{+} and γ_S^{-}) are obtained.

In the following section, the specimen calculation for workpieces before finishing is shown. For example, Conditions – Workpiece (Before finish); No. of cycles: 0; Parallel direction (X-axis). The measured contact angle: $\theta_W = 103.6°$; $\theta_F = 85.1°$; $\theta_D = 57.4°$. Surface tension components (γ^{LW}, $\gamma^{+}$and γ^{-}) are presented in Table 5.3 for different liquids. These values are substituted in the equation to calculate the surface energy.

$$[(1 + \cos 103.6) \times 72.8] = 2\left[(\gamma_s^{LW} \times 21.8)^{0.5} + (\gamma_s^{+} \times 25.5)^{0.5}\right.$$
$$\left. + (\gamma_s^{-} \times 25.5)^{0.5}\right] \quad (5.4)$$

$$[(1 + \cos 85.10) \times 58.0] = 2\left[(\gamma_s^{LW} \times 39.0)^{0.5} + (\gamma_s^{+} \times 2.28)^{0.5}\right.$$
$$\left. + (\gamma_s^{-} \times 39.6)^{0.5}\right] \quad (5.5)$$

$$[(1 + \cos 57.40) \times 50.8] = 2\left[(\gamma_s^{LW} \times 50.8)^{0.5} + (\gamma_s^{+} \times 0.10)^{0.5}\right.$$
$$\left. + (\gamma_s^{-} \times 0)^{0.5}\right] \quad (5.6)$$

Surface energy components: $\gamma_S^{LW}= 31.7$ mJ/m^2, $\gamma_S^{+}= 0.5$ mJ/m^2 and $\gamma_S^{-}= 1.4$ mJ/m^2

The surface free energy components obtained for SS316L materials for varying process parameters are presented in Figure 5.10. It depicts that for all the finished workpieces, the surface energy components are reduced compared to the unmodified surface. It may be due to the reduced surface area. The least surface energy of 27.8 mJ/m^2 is achieved on the surface finished with 40 bar pressure and 9 cycles, whereas unmodified surface energy is 40.1 mJ/m^2 with respect to the perpendicular viewing direction. Figure 5.11 shows the surface free energy components for Ti-6Al-4V materials. The result obtained shows a different trend in parallel and perpendicular directions to the surface. For many workpieces in the parallel direction, surface energy increases due to a decrease in contact angle, and in the perpendicular direction, contact angle increases hence surface energy decreases. Surface energy

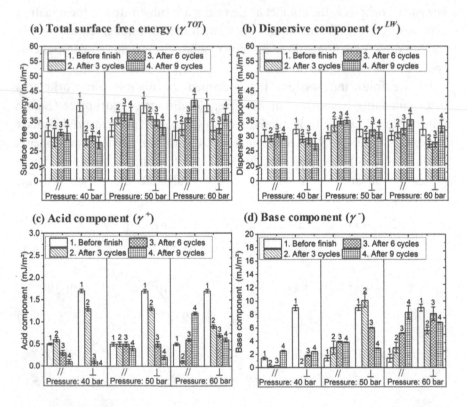

FIGURE 5.10 Variation in surface energy – SS316L. (a) Total surface free energy (γ^{TOT}), (b) Dispersive component (γ^{LW}), (c) Acid component (γ^+) and (d) Base component (γ^-).

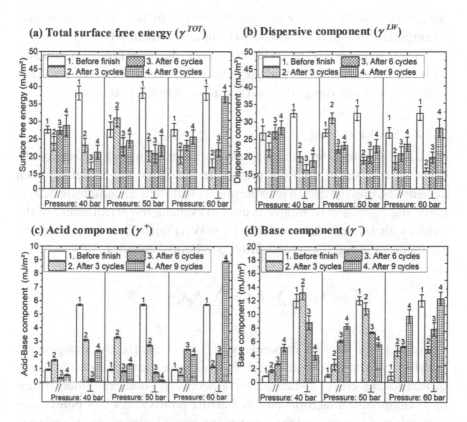

FIGURE 5.11 Variation in surface energy – Ti-6Al-4V. (a) Total surface free energy (γ^{TOT}), (b) Dispersive component (γ^{LW}), (c) Acid component (γ^+) and (d) Base component (γ^-).

follows the trend observed in the contact angle measurement for different liquids. Surface energy for some surfaces is almost close to each other for evaluated surfaces in a parallel direction.

A study conducted by Yan et al. (2017) on Ti-6Al-4V with the same liquids concluded that there is no clear functional relationship between the total surface energy when surface roughness varies in the range of 0.280 μm and 0.095 μm. The workpiece surfaces finished with the developed technique exhibit significantly less surface energy compared to the liquid surface tension. From this study, it is clear that there is a different tendency of the droplet contact angle and surface energy along the direction parallel and perpendicular to the finishing process observed for both the materials – SS316L and Ti-6Al-4V. It can be concluded that the surface roughness and the surface morphology play an important role in anisotropic wetting of the materials.

Further, an effort has been made to optimize the input process parameters such as a number of cycles and pressure to achieve the deserved responses such as R_a and MR using RSM and is explained in the next section.

5.3.4 RSM Modelling

RSM has been used for obtaining the optimum values of surface roughness (R_a) and MR of the work surface. In this work, the model has been developed using the design of experiments with central composite design. Later, the regression technique is used to evaluate the performance of the model with an analysis of variance (ANOVA) tests. The obtained results are modelled using the second-order polynomial equation 5.7:

$$ Y = \beta_0 + \sum_{i=1}^{k}\beta_i x_i + \sum_{i=1}^{k}\beta_{ii} x_i^2 + \sum_{i<j}^{k}\beta_{ij} x_i x_j \qquad (5.7) $$

where Y is the desired response; $x_i = 1, 2 \ldots k$ are the independent variables; β_0 is the constant; and β_i, β_{ii} and β_{ij} are the coefficients of linear, quadratics and cross product terms, respectively. Total experimental runs of 13 comprising 4 factorial points, 5 centre points and 4-star points have been selected to carry out the experimentation. Centre points are selected to get an idea of pure error, while star points are added to establish the relationship between independent variables and response. The number of cycles and pressure are selected as input parameters, and they are denoted as A and B, respectively. The R_a and MR are the output responses. The factors and levels are selected based on the preliminary experiments conducted on the machining of various components. An experimental setup has a constraint that the maximum system pressure of the hydraulic power pack is 100 bar, and it is not advisable to run the power pack at maximum capacity. To overcome this, maximum pressure of 60 bar is fixed in the present investigation. Second, due to the large quantity of the media used in this process, the surface may reach optimal surface roughness value with a fewer number of cycles. Further, it may damage the surface with the same abrasive mesh size. To overcome this also, the numbers of cycles are restricted to 9. Table 5.4 shows the layout of central composite design with responses R_a and MR.

TABLE 5.4   Layout of central composite design with responses

	Factor 1 A: No. of cycles [3, 6, 9]	Factor 2 B: Pressure (bar) [40, 50, 60]	Materials			
			SS316L		Ti-6Al-4V	
Std. Run			Response 1 R_a (μm)	Response 2 MR (mg)	Response 1 R_a (μm)	Response 2 MR (mg)
1	3	40	0.193	2.5	0.267	1.86
2	9	40	0.133	6.55	0.197	3.5
3	3	60	0.160	5.38	0.219	3.39
4	9	60	0.114	10.98	0.147	6.95
5	3	50	0.174	3.315	0.268	2.51
6	9	50	0.130	9.15	0.171	5.02
7	6	40	0.145	4.71	0.220	3.12
8	6	60	0.129	9.355	0.160	4.5
9	6	50	0.147	8.195	0.206	3.93
10	6	50	0.152	8.254	0.197	3.88
11	6	50	0.148	8.5	0.208	3.81
12	6	50	0.155	8.98	0.202	3.96
13	6	50	0.151	8.124	0.203	3.42

Further, the ANOVA technique is applied to characterize the performance of the RSM model. Table 5.5 presents the ANOVA table for R_a and MR for SS316L and Ti-6Al-4V materials. The F-value of the models 34.56, 38.87, 60.54 and 39.6 indicates that the model is significant for all the responses. The p-value of less than 0.05 indicates that the model terms are significant at 95% confidence level, as shown in the literature. The percentage contribution of each parameter towards surface roughness and MR is also presented in Table 5.5.

The following equations are used to make further predictions of the output responses for the given input levels of each factor.

$$R_a = 0.149 + (-0.025 \times A) + (-0.0113 \times B) + (0.0035 \times AB) \\ + (0.007 \times A^2) + (-0.0088 \times B^2) \tag{5.8}$$

$$MR = 8.20 + (2.58 \times A) + (1.99 \times B) + (0.38 \times AB) \\ + (-1.45 \times A^2) + (-0.65 \times B^2) \tag{5.9}$$

TABLE 5.5   ANOVA table for average surface roughness and material removed

Material/Responses	Source	Model	A-Cycles	B-Pressure	AB	A²	B²	Residual	Lack of Fit	Error	Total
SS316L material	Average surface roughness (Ra)										
	Sum of Squares	0.0048	0.0038	0.0008	0.0	0.0001	0.0002	0.0002	0.0002	0.0	0.0050
	F-value	34.56	135.08	27.76	1.77	4.87	6.37		4.96		
	p-value	<0.0001	<0.0001	0.0012	0.2257	0.063	0.0396		0.0781		
	% Contribution	96	76	16	0	2	4	4	4	0	100
	$R^2 = 0.9611$; Adjusted $R^2 = 0.9333$; Predicted $R^2 = 0.7382$; Adeq. Precision = 20.3005										
	Material removed (MR)										
	Sum of squares	74.95	39.96	23.82	0.6006	5.85	1.18	2.70	2.21	0.4854	77.65
	F-value	38.87	103.62	61.76	1.56	15.16	3.07		6.08		
	p-value	<0.0001	<0.0001	0.0001	0.2522	0.006	0.1232		0.0568		
	% Contribution	96.52	51.46	30.67	0.77	7.53	1.51	3.47	2.84	0.62	100
	$R^2 = 0.9652$; Adjusted $R^2 = 0.9404$; Predicted $R^2 = 0.7676$; Adeq. Precision = 21.6788										
Ti-6Al-4V material	Average surface roughness (Ra)										
	Sum of squares	0.146	0.0095	0.0042	1×10⁻⁶	0.0008	0.0004	0.0003	0.0003	0.0001	.0149
	F-value	60.54	197.48	86.31	0.0207	16.32	9.13		5.02		
	p-value	<0.0001	<0.0001	<0.0001	0.8895	0.0049	0.0194		0.0765		
	% Contribution	97.73	63.76	27.86	0.006	5.27	2.946	2.260	1.786	0.47	100
	$R^2 = 0.9774$; Adjusted $R^2 = 0.9613$; Predicted $R^2 = 0.8115$; Adeq. Precision = 28.0546										
	Material removed (MR)										
	Sum of squares	17.6	9.91	6.74	0.9216	0.0039	0.0188	0.6223	0.4289	0.1934	18.23
	F-value	39.6	111.44	75.83	10.37	0.0435	0.211		2.96		
	p-value	<0.0001	<0.0001	<0.0001	0.0147	0.8407	0.6599		0.1611		
	% Contribution	96.54	54.36	36.97	5.055	0.021	0.103	3.413	2.352	0.62	100
	$R^2 = 0.9659$; Adjusted $R^2 = 0.9415$; Predicted $R^2 = 0.7478$; Adeq. Precision = 23.1533										

$$R_a = 0.2030 + (-0.0398 \times A) + (-0.0263 \times B)$$
$$+ (-0.0005 \times AB) + (0.0169 \times A^2) + (-0.0126 \times B^2) \quad (5.10)$$

$$MR = 3.78 + (1.28 \times A) + (1.06 \times B) + (0.48 \times AB)$$
$$+ (0.037 \times A^2) + (0.082 \times B^2) \quad (5.11)$$

R_a and MR – SS316L is given by:

R_a and MR – Ti-6Al-4V

Another important factor is the correlation coefficient (R^2), and it should be close to unity for a model to be accurate. The R^2 value is found to be near unity for all the models. Also, the predicted R^2 is in the reasonable agreement with the adjusted R^2; that is, the difference is less than 0.2.

The interaction of process parameters such as number of cycles and pressure on R_a and MR is plotted using the RSM surface plot. Figure 5.12 shows the three-dimensional (3D) response surface plot for surface roughness and MR for different materials. This depicts that, as the number of cycles and pressure increases, surface roughness decreases, and MR increases irrespective of the type of material. The variation in R_a is proportional to a number of cycles, and pressure has less influence on R_a. Similarly, an exponential variation of MR is observed with a given number of cycles and pressure. Also, the 3D response plots can be used for prediction of the responses at different levels of input factors. In case of SS316L, a linear variation in surface roughness is observed, and also the effect of a number of cycles on R_a (percentage contribution – no. of cycles: 76% and -pressure: 16% – Ref. Table 5.5) and MR (percentage contribution – no. of cycles: 51.46% and pressure: 30.67% – Ref. Table 5.5) is more. Also, it is observed from the surface plot, after 6 cycles and 55 bar pressure, the MR is less as compared to initial cycles. For Ti-6Al-4V, 3D response surface plots clearly depict the nonlinear variation in R_a and linear variation in MR.

From the parametric study, it is observed that the percentage contribution of the number of cycles in finishing of both the materials is more compared to pressure. Based on the above results, in the next level of experiments, the number of cycles and abrasive media size is varied, and experiments are carried out. Further, these surfaces are checked for biocompatibility, and results are explained in the next section.

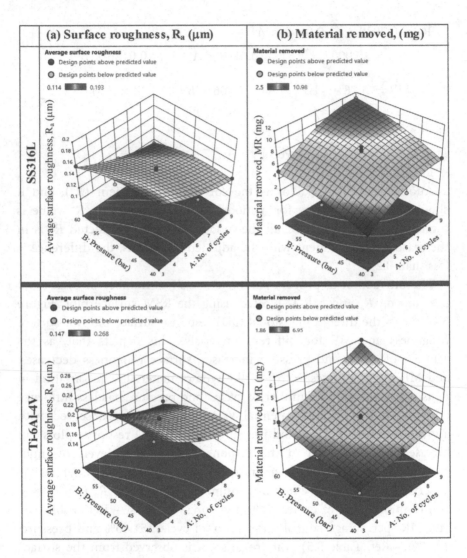

FIGURE 5.12   3D response surface plots. (a) Surface roughness, R_a (µm) and (b) material removed (mg).

5.4 EFFECT OF SURFACE ROUGHNESS ON BACTERIAL ADHESION STUDY

This experimental study aims to understand the influence of finishing parameters such as a number of cycles and abrasive particle size on the surface roughness, surface morphology and bacterial adhesion. The material selection and fixture development for these experiments are presented in Chapter 4. Experimental details are presented in Table 5.6. For each experiment, the individual samples are considered. Procedure for bacterial adhesion study and results of finishing and bacterial adhesion study are explained in this section.

5.4.1 Procedure for Bacterial Adhesion Study

Figure 5.13 (a) shows the procedural steps followed in bacterial adhesion study. For the bacterial adhesion assessment, all the samples – SS316L and Ti-6Al-4V – finished with different process parameters are placed in an individual well (place to hold the samples) in 24-well plates. For each study, three independent samples are considered for counting bacterial adhesion, and one sample is used for SEM analysis

TABLE 5.6   Experimental details for bacterial adhesion study

Parameters	Particulars			
Sample materials	(a) ASTM F138: SS316L			
	(b) ASTM F136: Ti-6Al-4V			
Sample size and shape	Circular disc: 13 mm diameter × 6 mm thickness			
Abrasive media	Type	Abrasive particles	Polymer	Plasticizer
	1	SiC mesh size 220	Silicone rubber	Silicone oil
	2	SiC mesh size 400	Silicone rubber	Silicone oil
Abrasive concentration	50% abrasive particles + 38% silicone rubber + 12% silicone oil			
Pressure	50 bar			
Stroke length	400 mm			

	Sample details		
Sample	Mesh size	No. of cycles	Coding
Sample 1	–	0	As-received
Sample 2	#220	3	#220 – 3C
Sample 3	#220	6	#220 – 6C
Sample 4	#400	3	#400 – 3C
Sample 5	#400	6	#400 – 6C
Sample 6	#220 and #400	3, 3	Hybrid

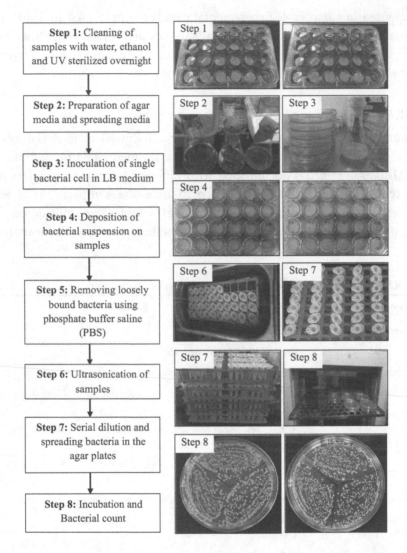

FIGURE 5.13 A   Procedural steps followed in bacterial adhesion study.

(total number of samples SS316L: 48 no. and Ti-6Al-4V: 48 no.). The step by step procedure is detailed below:

- **Step 1 – Washing of the samples:** The samples are washed with water, ethanol and UV sterilized overnight to remove the contamination over the surfaces. Two different bacterial strains *Escherichia coli* referred to as *E. coli* (ATCC 25922) and *Staphylococcus aureus,* referred to as *S. aureus* (ATCC 25923) is used for this study because these bacterial strains are commonly found in orthopaedic implant-related infections.

- **Step 2 – Preparation and spreading of agar media:** Lysogeny Broth (LB) medium is a nutrient-rich media, which is commonly used to culture bacteria, is prepared, and spread over the plates for the study.

- **Step 3 – Inoculation:** A single colony of the culture is inoculated into 50 ml LB medium and incubated at 37°C at 180 rpm for growth.

- **Step 4 – Deposition:** After the culture reached the logarithmic growth phase (Optical Density ≈ 0.4) confirmed with a spectrophotometer, 2 ml of the bacterial suspension is transferred into sample well plates and incubated for 12 h at 37°C.

- **Step 5 – Removing loosely bound bacteria:** After incubation, the samples are carefully dipped in sterile phosphate buffer saline (PBS) twice to remove loosely bound bacteria and transferred to another sterile plate.

- **Step 6 – Ultrasonication:** Then, the samples are sonicated in fresh LB broth to remove the adhered bacteria.

- **Step 7 – Serial dilution and spreading:** Subsequently, the culture is serially diluted to reduce the concentration and spread plated on the LB agar plate for the growth of bacteria.

- **Step 8 – Incubation and bacterial count:** After overnight incubation at 37°C, the viable colonies are counted to calculate the colony-forming units (CFUs) and expressed as a number of bacteria adhered to samples per mm^2 (CFU/mm^2).

The adhered bacteria on the surfaces are observed using HRSEM, and the procedure to prepare the samples for SEM is shown in Figure 5.13 (b). Briefly, after PBS wash (to remove any nonadherent bacteria), the bacterial adhered samples are incubated in 2.5% glutaraldehyde for 2 h at 37°C to fix the cells. Then, the samples are dehydrated with a series of ethanol concentrations in water (50%, 70%, 90% and 100% vol.) for 15 min each to remove the water content from the samples. After complete

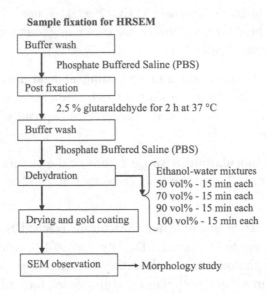

FIGURE 5.13 B   Sample preparation procedure for HRSEM.

dehydration, the samples are observed under SEM for analysis on bacterial adhesion. Statistical significant differences between as-received samples and finished surfaces are analysed using one-way ANOVA in GraphPad Prism version 8 software. The experimental data are represented as a mean ± standard deviation with a number of replicates. Statistical significance is represented by p values, a value of *p < 0.05 is considered significant, a value of **p < 0.01 is considered very significant, values of ***p < 0.001 and ****p < 0.0001 are considered extremely significant.

5.4.2 Finishing of Biomaterials for Bacterial Adhesion Study

Figure 5.14 (a) shows a change in the R_a of the SS316L individual samples with a different number of cycles and abrasive media. Initially, the $R_a R_a$ of the as-received samples is 0.463 μm. After finishing with abrasive media of 220 mesh size at 3 cycles, R_a is reduced to 0.163 μm, and in another 3 cycles, it is further reduced to 0.113 μm. For 400 mesh abrasive media, initial R_a from 0.463 μm reduced to 0.213 μm at 3 cycles and further reduced to 0.139 μm roughness, and significant differences in change in R_a are observed for different samples.

Figure 5.14 (b) shows a change in the R_a of the Ti-6Al-4V with a different number of cycles and abrasive media. The R_a of the as-received sample is 0.402 μm. After finishing with abrasive media of 220 mesh size at

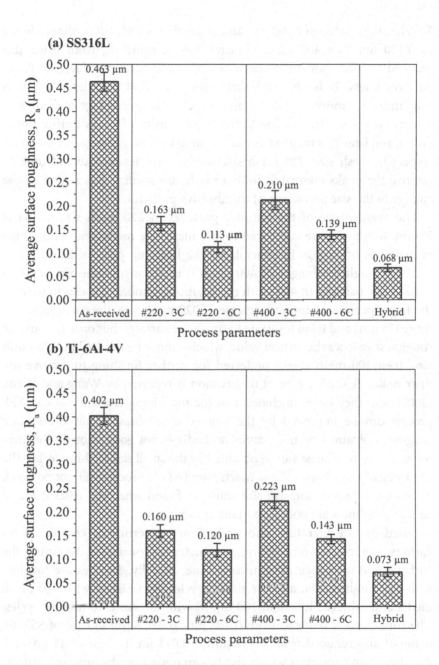

FIGURE 5.14 Change in the average surface roughness with different process parameters. (a) SS316L and (b) Ti-6Al-4V.

3 cycles, R_a reduced to 0.160 μm, and in another 3 cycles, it further reduced to 0.120 μm. For 400 mesh abrasive media, initial R_a from 0.463 μm reduced to 0.223 μm roughness at 3 cycles and further reduced to 0.143 μm roughness. It is observed that during the first 3 cycles, maximum roughness is removed, while in later 3 cycles, the percentage of change in roughness is less. This is due to the large density of peaks during initial cycles, and later it is reduced as the finishing process progresses. Also, with respect to mesh size, 220 mesh size media takes less amount of time to remove the peaks compared to 400 mesh size media. This is due to the change in the size and shape of the abrasive particles.

The average size of the abrasive particles in 220 mesh size media is 75 μm, which is more compared to 400 mesh size media with an average particle size of 30 μm. In general, for the finishing process, 220 mesh is used for rough finishing, and 400 mesh is used for fine finishing. Due to smaller size particles in 400 mesh media, more number of cycles to remove the larger peaks are required. In case of 220 mesh size, the particles size is more (75 μm) and takes less time to finish the surfaces, but once the surface roughness reaches the critical value, which cannot be finished by 220 mesh size, then 400 mesh size is preferred for further finishing to remove the finer peaks. A similar type of observation is reported by Wang and Weng (2007), and they have concluded that the machining efficiency of the AFF process can be improved by the large size of abrasive, but the surface roughness obtained by these abrasive media is not good enough. Furthermore, a finer roughness can be obtained by the small size of abrasive, but the efficiency of the Abrasive Flow Machining (AFM) process will not be good. Therefore, a proper abrasive size must be found when the efficiency and surface roughness are both important factors.

Based on the literature and preliminary experiments conducted on various workpieces, one hybrid parameter is considered to finish the surfaces and to reduce the finishing time. Initially, 3 cycles are finished with 220 mesh media, and the next 3 cycles are finished with 400 mesh media and achieved less than 100 nm R_a with the same number of cycles. The roughness is reduced from 0.463 μm to 0.068 μm in case of SS316L material and reduced from 0.402 μm to 0.073 μm in case of Ti-6Al-4V. Another main parameter which also has an impact on the finishing process is material hardness because it is observed from the results that the percentage change in roughness is more in SS316L (84% in case of hybrid parameters) compared to Ti-6Al-4V (81% in case of hybrid parameters) because the hardness of the SS316L is less compared to Ti-6Al-4V material. Figure 5.15 shows the profilometer images showing the work surface lay patterns with different finishing conditions. It shows the variation in the height of peaks and valleys with respect to changes in the surface roughness value. The peak height variations are more in received samples, and as the finishing process progresses, the variation is reduced for both the materials.

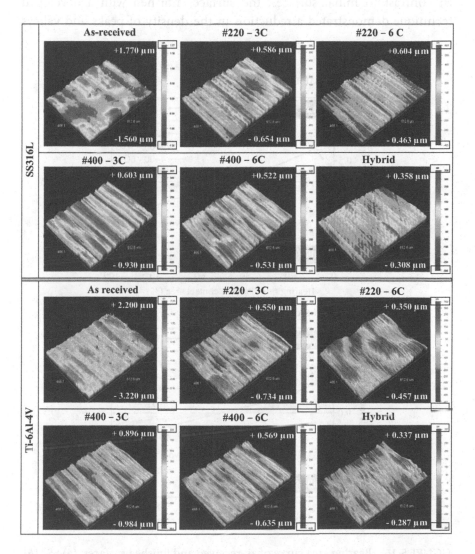

FIGURE 5.15 Profilometer images showing the work surface lay patterns with different finishing conditions.

In the present work, an attempt has been made to show graphically BAC of SS316L and BAC of Ti-6Al-4V before and after finishing. In the case of the implants, the bearing area of the surface may be correlated with the biomechanical properties of the surface and the long-term anchoring of the implant (Kanthababu et al., 2009; Loberg et al., 2010; Laheurte et al., 2012). Figure 5.16 shows the BAC for the as-received sample, and samples finished with different abrasive media and cycles. In contrast to initial surfaces, the surfaces finished with a developed technique demonstrated a reduction in the density of peaks and valleys

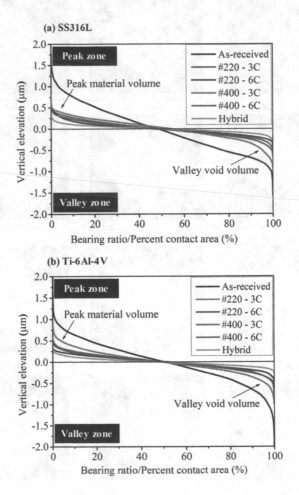

FIGURE 5.16   Bearing area curve of as-received and finished surfaces. (a) SS316L and (b) Ti-6Al-4V.

irrespective of the materials. Surface asperity – peaks of the surface finished with hybrid parameters – is less compared to other surfaces, and it got fewer irregularities. Further, these finished surfaces are subjected to bacterial adhesion study to check the biocompatibility of the surfaces, and results are presented in the next section.

5.4.3 Bacterial Adhesion

Initial bacterial adhesion and biofilm formation is a multi-factor phenomenon, as discussed in the literature survey, and it is very difficult to conclude with few factors alone. Although the experimental result shows that the reduced bacterial adhesion is observed on the finished surface, several limitations must be noted when interpreting the results. There are other factors like bacterial surface properties, and surrounding environmental factors might also have an influence on the bacterial adhesion.

The main objective of this work is to characterize and identify the process parameters to finish biomaterials using flexible abrasive media and to study its effect on bacterial adhesion. This work provides valuable insights on how the proposed finishing process can reduce the surface roughness and its effect on bacterial adhesion and the process adaptability for finishing the orthopaedic implants. However, the limitations such as a selection of abrasive media, design of fixture (based on the shape of the particular implant), surface cleaning of the device (after finishing process) and monitoring of the abrasive media are the challenges yet to overcome.

In the present work, it is hypothesized that the decreased surface roughness by abrasive flow finishing will lead to reduced bacterial adhesion. Therefore, the effect of physical characteristics, such as surface roughness and surface morphology of the material surface on bacterial adhesion, is explored. Figure 5.17 shows the adhesion of E. coli on SS316L and Ti-6AL-4V surfaces. For SS316L, the as-received sample has a bacterial count of 4.99×10^{10} CFU/mm^2 (Figure 5.17 a). The decrease in surface roughness by varying finishing parameters tends to reduce bacterial adhesion. For hybrid parameter finishing, there is a significant reduction in a number of bacterial colonies (6.86×10^9 CFU/mm^2).

A similar trend in bacterial reduction is observed in the case of Ti-6Al-4V as well (Figure 5.17 b). The colonies reduced from 3.13×10^{10} CFU/mm^2 to 6.53×10^8 CFU/mm^2 as the roughness decreased for the hybrid

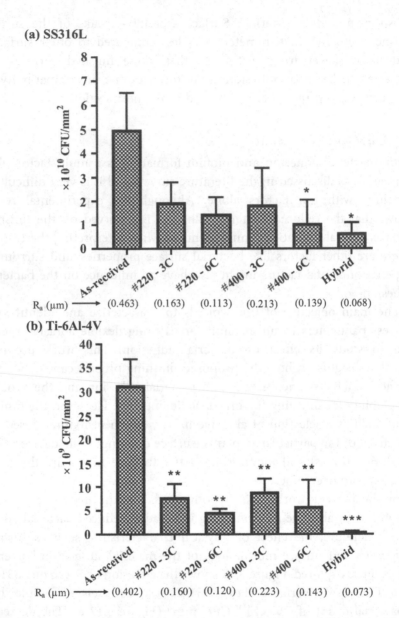

FIGURE 5.17 *E. coli* adhesion on samples (*p < 0.05, **p < 0.01, ***p < 0.001 vs. as-received).(a) SS316L and (b) Ti-6Al-4V.

finished sample. Our data indicate that the SS316L sample used is more susceptible to bacterial adhesion compared to the Ti-6Al-4V sample. The possible reason might be that the initial roughness range of titanium samples is less compared to the stainless steel samples. A similar trend is observed in S. aureus adhesion on both the samples (Figure 5.18). The adhesion of S. aureus on SS316L surfaces reduced from 1.39×10^{11} CFU/mm^2 to 3.41×10^{10} CFU/mm^2 after hybrid finishing (Figure 5.18 a). Also, for the same roughness value, Ti-6Al-4V samples show less bacterial adhesion because it has good antibacterial properties compared to SS316L.

Similarly, Figure 5.18 (b) shows the adhesion of S. aureus on Ti-6Al-4V reduced from 1.28×10^{11} CFU/mm^2 to 1.27×10^{10} CFU/mm^2. For both the materials, less bacterial adhesion is observed on the surfaces finished with hybrid parameters (it can also be called as multi-stage finishing). The results depict that there are statistically significant differences in the total number of bacteria adhered to SS316L and Ti-6Al-4V between as-received and finished surfaces. This may be due to reduced surface roughness and increased hydrophobicity of the finished surfaces.

Hauslich et al. (2013) reported that the increased adhesion onto rough surfaces might be due to enlarged surface areas related to irregularities created during the manufacture of these surfaces. The quantity of E. coli adhesion on other finished surfaces is almost similar because the roughness values (R_a) of these surfaces are all in the range of below 200 nm. A similar observation is made in a study conducted by Quirynen and Bollen (2005) and Wassmann et al. (2017). The surface roughness of 0.200 μm R_a is the threshold R_a below which the roughness does not have an effect on the bacterial adhesion. But for S. aureus, such correlations are not observed because of the significant variation in the bacterial adhesion is observed among the other finished surfaces.

The colony count method is further supported by HRSEM images of bacterial adhesion on the material surfaces (Figure 5.19). Adhesion of both the bacteria E. coli and S. aureus decreased on finished surfaces compared to the as-received surfaces. From the HRSEM images (Figure 4.35) of the as-received sample of SS316L, it is observed that S. aureus bacteria are found to be localized in the few crevices present on the surface.

The amount of the bacterial colonies on the finished surfaces is less compared to the as-received surfaces. On some of the finished surfaces, E. coli bacteria lay across the groove direction, this may be

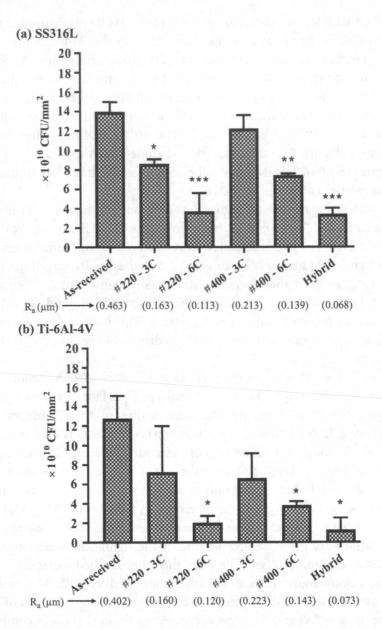

FIGURE 5.18   *S. aureus* adhesion on samples (*p < 0.05, **p < 0.01, ***p < 0.001 vs. as-received). (a) SS316L and (b) Ti-6Al-4V.

because the width and depth of the features are significantly smaller than the cells, and similar type of observations is reported by Edwards and Rutenberg (2001) and Mortimer et al. (2016). Further, Figure 5.20

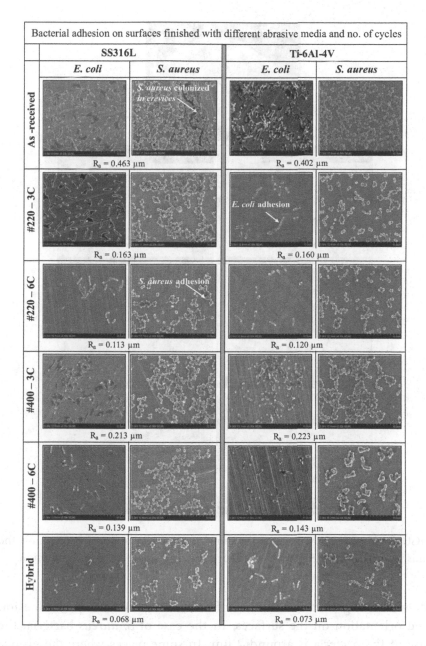

FIGURE 5.19   HRSEM images of bacterial adhesion on samples.

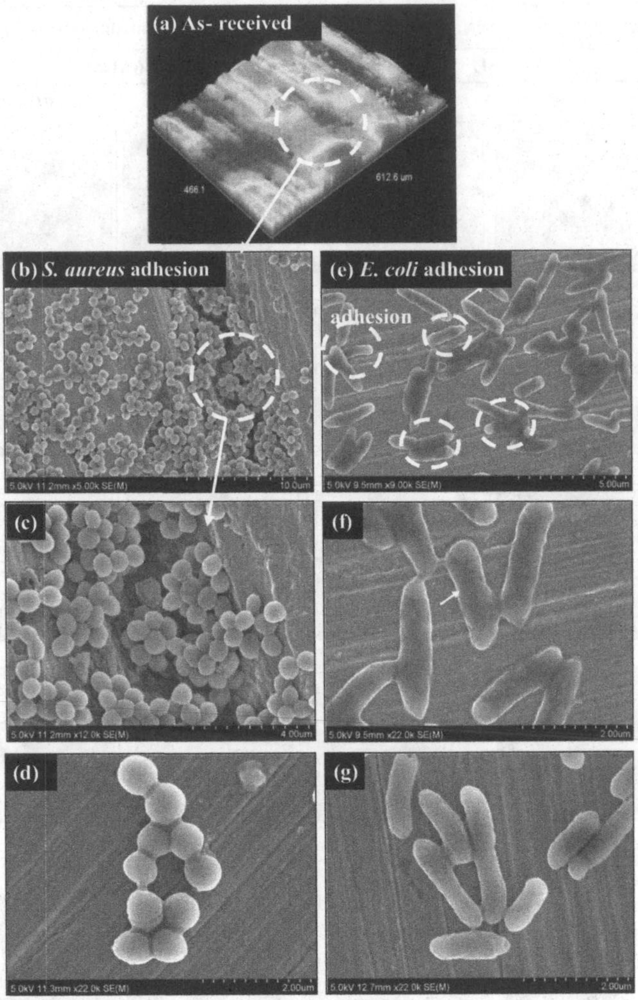

FIGURE 5.20   HRSEM images are showing the orientation of bacteria on the finished surfaces. (a) As-received, (b) *S. aureus* adhesion, (e) *E. coli* adhesion.

shows the higher magnification of SEM images of bacterial adhesion and orientation on the surfaces. In the case of *S. aureus* bacteria, the size of the bacteria is around 2 μm. In some places where the groove width is more than the bacterial size, *S. aureus* bacteria localized in the grooves-like clusters can be observed in Figure 5.20 (b-d). In the case of *E. coli* also, on as-received surfaces, some places where the width of the groove is more, the bacteria is oriented toward the direction of the groove. But in some places where the groove size is smaller, the bacteria are oriented in the perpendicular direction observed in Figure 5.20 (e.g., highlighted in circle and arrows).

The possible reason may be binding energy for perpendicular alignment is larger than for longitudinal alignment along small grooves where the groove sizes are less than bacterial size, which are in qualitative agreement with the experimental observations where bacteria seem to align perpendicular to small grooves. This phenomenon is clearly explained in the study conducted by Edwards and Rutenberg (2001). Studies also show that the bacteria are more susceptible to the rough surface, and it tends to accumulate more in concave features due to the increased contact area (Yoda et al., 2014).

Morphological description of the surface patterns such as grooved surface, braided surface, porous surface and scratches have more influence on initial bacterial adhesion along with roughness parameters (Edwards and Rutenberg, 2001; Katsikogianni and Missirlis, 2004; Ribeiro et al., 2012). Even though the grooves are present on both the unmodified and modified surfaces, further analysis is to be carried out to identify the reasons for the change in densities of bacterial adhesion.

Figure 5.21 shows the two-dimensional roughness profile of the as-received sample and UAFF finished sample along X-axis and Y-axis. The X-axis shows the roughness profile perpendicular to the grooves, and the Y-axis shows the roughness profile along the grooves. It is observed that for both the SS316L and the Ti-6AL-4V, as-received samples have few grooves with larger height and width (20 μm to 40 μm) and also appeared to be significantly irregular than the hybrid samples with smoother and smaller grooves. It could be that these smaller sized grooves could not possibly accommodate bacteria which have considerably larger dimensions, and bacterial colonies can be filled more in the grooves with a higher width (Edwards and Rutenberg, 2001; Barbour et al., 2007; Wassmann et al., 2017).

It is also corroborated by the BAC (Figure 5.16), which shows that the volume of the peaks and valleys of as-received samples is higher than the finished samples. Having a low volume of peaks and valleys reduce the accommodation of bacterial cells, while the higher volume on as-received samples favours the bacterial accommodation. The reduced bacterial adhesion on the finished surfaces shows that the developed process can be used to finish the orthopaedic implants, and

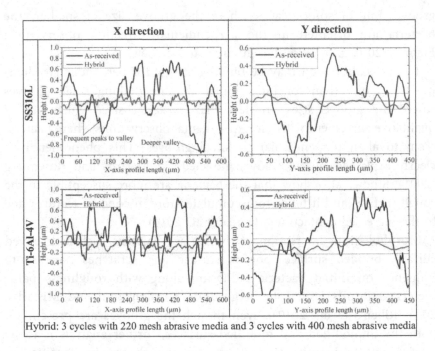

FIGURE 5.21 Two-dimensional roughness profiles – as-received sample and UAFF finished sample.

it would reduce the PII, but the clinical verification (i.e., *In Vivo*) would eventually be needed to check the compatibility.

5.5 SUMMARY

The machinability experiments are carried out on four different materials having different hardness numbers to understand the effect of a number of cycles on finishing of different materials. Effect of a number of cycles and pressure on surface roughness and surface morphology of the biomaterials are studied, and further, these work surfaces are subjected to wettability study. Similarly, the biomaterials are finished with varying number of cycles and different abrasive media sizes, and the obtained surface roughness and surface morphology of the work surface are examined. After finishing, these surfaces are subjected to bacterial adhesion study to understand the effect of surface roughness on bacterial adhesion. The major conclusions drawn based on the experimentation carried out in the present work are listed in Chapter 6.

CHAPTER 6

Summary and Conclusions

T his book gives complete information on the bio-implants-related infections, manufacturing difficulties in machining the bio-implants, and sequence of steps involved in the present invention, that is, development of the advanced finishing process to finish the bio-implants. In the current research work to overcome the challenges associated with conventional finishing processes to finish complex internal and external surfaces and biomaterials, a unidirectional abrasive flow finishing (UAFF) process is developed. The critical parameters in achieving the required surface roughness are studied. The essential parameters of interest are the number of cycles, pressure and abrasive media, which can be varied at different levels to study their effect on the output responses. The experimental setup has been developed in three stages, such as screw type abrasive flow finishing, unidirectional abrasive flow machining and closed loop unidirectional abrasive flow machining. The main components, along with other subsystems, are selected carefully based on the sample design parameters.

Simultaneously, viscoelastic polymer-based abrasive media has been developed with varying sizes of abrasive mesh size for the finishing of various surfaces. The developed abrasive media is subjected to different characterization studies to ascertain the thermal and rheological properties of the abrasive media. Further using the developed experimental setup and abrasive media, four sets of experiments are carried out: (1) machinability study has been carried out on different workpiece materials having different hardness values to study the effect of the number of cycles on surface roughness; (2) wettability study on machined biomaterials to investigate the influence of process parameters on surface roughness and contact angleand (3) bacterial adhesion study on machined biomaterials to investigate the influence of process parameters on surface roughness and bacterial adhesion. The obtained results are reported systematically in the thesis in different sections and chapters. The conclusions based on the experimentations are explained in the next section.

The significant findings arrived based on the extensive experiments carried out in the present research work are as follows:

(a) **Abrasive media development and characterization**

Viscoelastic polymer-based flexible abrasive media is developed. The developed abrasive media is subjected to thermal and rheological characterization, and obtained results are listed below:

- Thermogravimetric analysis results show that the maximum degradation temperature of the abrasive media is 576°C and the media can perform better at lower temperatures less than 100°C.

- Rheological studies carried out on the developed media show that the media exhibits the highest viscosity 4.5×10^5 Pa.s at a lower shear rate of 0.01 s^{-1}, and it is reduced to 69.7 Pa.s at a higher shear rate of 100 s^{-1}. As the shear rate increases, the media viscosity decreases due to the weak molecular force of attraction, causing the easy flow of media. Abrasive media shows the time-dependent shear thinning behaviour.

- The obtained storage modulus (G') and loss modulus (G'') reveal that the developed media behaves like an elastic solid (G' > G'') at different temperatures which are suitable for finishing the complex internal and the external features efficiently and effectively.

(b) **Machinability study**

Machinability study has been carried out on four different materials – aluminium, brass, copper and mild steel with varying number of cycles. The obtained results show that:

- Significant percentage improvement in average surface roughness (R_a)value is observed: 59.62% for aluminium, 56.12% for brass, 70.52% for copper and 46.53% for mild steel, respectively. It reveals that the softer materials are achieving the required finish with less number of cycles than harder materials.

- Scanning electron microscopic images show that there is a significant reduction in the deep machining feed marks left on the finished surfaces in conventional machining after 12 cycles of this finishing process.

(c) **Effect of surface roughness on wettability**

The effect of surface roughness on the wettability of the machined surfaces is observed on SS316L and Ti-6Al-4V. These materials are finished with a varying number of cycle and pressure. The obtained results are as follows:

- The minimum R_a of 0.114 μm is achieved on SS316L surfaces after finishing with 9 cycles with 60 bar pressure. Similarly, for Ti-6Al-4V, the R_a is reduced from 0.350 μm to 0.147 μm for the same process parameters.

  Wettability study depicts that the surface roughness and surface morphology have a great influence on the spreading of fluid on the surfaces. Spreading of droplets takes place always along the direction of the grooves. The change in contact angle is shown below:

SS316L	Contact angle (°)		Ti-6Al-4V	Contact angle (°)	
	$\theta^{\parallel}$	$\theta^{\perp}$		$\theta^{\parallel}$	$\theta^{\perp}$
Before	103.6	92.2	Before	98	84.5
After	89.5	90.3	After	89.3	90.6

- The machined surfaces also show that the contact angle and the surface energy along and across the direction of lay patterns are different because of alteration in surface morphology before and after finishing.

(d) **Effect of surface roughness on the bacterial adhesion**

An attempt has been made to finish the SS316L and the Ti-6Al-4V samples with different abrasive media and the number of cycles to study the effect of surface roughness on bacterial adhesion.

- The R_a of the as-received sample is 0.463 μm, and after finishing with abrasive media of 220 mesh size at 3 cycles, it is reduced to 0.163 μm, and in another 3 cycles, again it reduced to 0.113 μm. A similar trend is observed for all the samples finished with different parameters.

- The minimum surface roughness of 0.068 μm on SS316L and 0.072 μm on Ti-6Al-4V is achieved on the surfaces finished with a hybrid parameter – 3 cycles of 220 mesh and 3 cycles of 400 mesh abrasive media.

  The results of a number of *E. coli* and *S. aureus* bacterial adhered on SS316L and Ti-6Al-4V as-received sample and hybrid samples are listed below:

SS316L	Bacteria adhesion (CFU/mm^2)		Ti-6Al-4V	Bacteria adhesion (CFU/mm^2)	
	E. coli	S. aureus		E. coli	S. aureus
Before	4.99×10^{10}	1.39×10^{11}	Before	3.13×10^{10}	1.28×10^{11}
After	6.86×10^{9}	3.41×10^{10}	After	6.53×10^{8}	1.27×10^{10}

- The reduced bacterial adhesion on the finished surface shows that the developed process can be used efficiently and effectively used to finish the bio-implants, and it would reduce the prosthetic implant infection, but clinical verification (i.e.*in vivo*) would eventually be needed to check the compatibility.

References

Aggarwal, S.L. (1987) *Rubber Technology*, 3rd Edition, M.Morton, Ed., Van Nostrand Reinhold, New York.

Aho, J., J.P.Boetker, S.Baldursdottir, and J.Rantanen (2015) Rheology as a tool for evaluation of melt processability of innovative dosage forms. *International Journal of Pharmaceutics*, **494**(2), 623–642.

Alam, F., V.Verma, and K.Balani (2015) Fundamentals of surface modification. *Biosurfaces: A Materials Science and Engineering Perspective*, 126–145. doi:https://doi.org/10.1002/9781118950623.ch4.

Ali-Tavoli, M., N.Nariman-Zadeh, A.Khakhali, and M.Mehran (2006) Multi-objective optimization of abrasive flow machining processes using polynomial neural networks and genetic algorithms. *Machining Science and Technology*, **10**(4), 491–510.

An, Y.H. and R.J.Friedman (2000) *Handbook of Bacterial Adhesion*, Humana Press Inc., Totowa, NJ.

Arciola, C.R., C.Campoccia, G.D.Ehrlich, and L.Montanaro (2015) Biofilm-based implant infections in orthopaedics. Biofilm-based healthcare-associated infections: volume I. *Advances in Experimental Medicine and Biology*, **830**, 29–46.

Arifvianto, B., Suyitno, and M.Mahardika (2012) Effects of surface mechanical attrition treatment (SMAT) on a rough surface of AISI 316L stainless steel. *Applied Surface Science*, **258**(10), 4538–4543.

Barbour, M.E., D.J.O'Sullivan, H.F.Jenkinson, and D.C.Jagger (2007) The effects of polishing methods on surface morphology, roughness and bacterial colonisation of titanium abutments. *Journal of Materials Science: Materials in Medicine*, **18**(7), 1439–1447.

Bhushan, B. (2013) *Introduction to Tribology*, 2nd Edition, John Wiley & Sons Ltd, The Atrium, West Sussex, United Kingdom.

Bikiaris, D. (2010) Microstructure and properties of polypropylene/carbon nanotube nanocomposites. *Materials*, **3**(4), 2884–2946.

Bio-Implants Market – by Type (Cardiovascular, Spine, Orthopedics, Trauma, Dental), by ROA (Surgical/Injectable), by Origin (Allo/Auto/Xenograft, Synthetic) &Materials (Ceramics, Biomaterial, Alloys, Polymers) – Global Trends &, Forecasts till 2017: weblink: www.marketsandmarkets.com/pdfdownloadNew.asp?id=728

Bohinc, K., G.Drazic, A.Abram, M.Jevsnik, B.Jersek, D.Nipic, M.Kurincic, and P.Raspor (2016) Metal surface characteristics dictate bacterial adhesion capacity. *International Journal of Adhesion and Adhesives*, **68**, 39–46.

Cassie, A.B.D. and S.Baxter (1944) Wettability of porous surfaces. *Transactions of the Faraday Society*, **40**, 546–551.

Chan, C.W., L.Carson, G.C.Smith, A.Morelli, and S.Lee (2017) Enhancing the antibacterial performance of orthopaedic implant materials by fibre laser surface engineering. *Applied Surface Science*, **404**, 67–81.

Chen, W.C., K.L.Wu, and B.H.Yan (2014) A study on the application of newly developed magneto-elastic abrasive to improving the surface roughness of the bore. *International Journal of Advanced Manufacturing Technology*, **73**(9–12), 1557–1566.

Cheng, C.T., G.Zhang, and S.To (2016) Wetting characteristics of bare micro-patterned cyclic olefin copolymer surfaces fabricated by ultra-precision raster milling. *RSC Advances*, **6**, 1562–1570.

Choudhury, D., M.Vrbka, A.B.Mamat, I.Stavness, C.K.Roy, R.Mootanah, and I.Krupka (2017) The impact of surface and geometry on coefficient of friction of artificial hip joints. *Journal of the Mechanical Behavior of Biomedical Materials*, **72**, 192–199.

Cox, S.C., P.Jamshidi, N.M.Eisenstein, M.A.Webber, H.Burton, R.J.A.Moakes, O.Addison, M.Attallah, D.E.T.Shepherd, and L.M.Grover (2017) Surface finish has a critical influence on biofilm formation and mammalian cell attachment to additively manufactured prosthetics. *ACS Biomaterials Science and Engineering*, **3**(8), 1616–1626.

Cross, M.M. (1979) Relation between viscoelasticity and shear-thinning behaviour in liquids. *RheologicaActa*, **18**(5), 609–614.

Cunha, A., A.P.Serro, V.Oliveira, A.Almeida, R.Vilar, and M.C.Durrieu (2013) Wetting behaviour of femtosecond laser textured Ti-6Al-4V surfaces. *Applied Surface Science*, **265**, 688–696.

Dabrowski, L., M.Marciniak, and T.Szewczyk (2006) Analysis of abrasive flow machining with an electrochemical process aid. *Proceedings of the Institution of Mechanical Engineers, Part B: Journal of Engineering Manufacture*, **220**, 397–403.

Das, M., V.K.Jain, and P.S.Ghoshdastidar (2012b) Computational fluid dynamics simulation and experimental investigations into the magnetic-field-assisted nano-finishing process. *Proceedings of the Institution of Mechanical Engineers, Part B: Journal of Engineering Manufacture*, **226**(7), 1143–1158.

Das, M., V.K.Jain, and P.S.Ghoshdastidar (2012a) Nanofinishing of flat workpieces using rotational – magnetorheological abrasive flow finishing (R-MRAFF) process. *International Journal of Advanced Manufacturing Technology*, **62**, 405–420.

Davies, P.J. and A.J.Fletcher (1995) The assessment of the rheological characteristics of various polyborosiloxane/grit mixtures as utilized in the abrasive flow machining process. *Proceedings of the Institution of Mechanical Engineers, Part C: Journal of Mechanical Engineering Science*, **209**(6), 409–418.

Edwards, K.J. and A.D.Rutenberg (2001) Microbial response to surface microtopography: the role of metabolism in localized mineral dissolution. *Chemical Geology*, **180**(1–4), 19–32.

Essabir, H., D.Rodrigue, R.Bouhfid, and A.E.K.Qaiss (2018) Effect of nylon 6 (PA6) addition on the properties of glass fiber reinforced acrylonitrile-butadiene-styrene. *Polymer Composites*, **39**(1), 14–21.

Fischer, G., M.Bigerelle, K.J.Kubiak, T.G.Mathia, Z.Khatir, and K.Anselme (2014) Wetting of anisotropic sinusoidal surfaces-experimental and numerical study of directional spreading. *Surface Topography: Metrology and Properties*, **2**(4), 044003.

Fu, Y., H.Gao, X.Wang, and D.Guo (2017) Machining the integral impeller and blisk of aero-engines: a review of surface finishing and strengthening technologies. *Chinese Journal of Mechanical Engineering (English Edition)*, 30(3), 528–543.

Gadelmawla, E.S., M.M.Koura, T.M.A.Maksoud, I.M.Elewa, and H.H.Soliman (2002) Roughness parameters. *Journal of Materials Processing Technology*, 123(1), 133–145.

Garg, H., G.Bedi, and A.Garg (2012) Implantsurface modifications: a review. *Journal of Clinical and Diagnostic Research*, 6(2), 319–324. doi:https://doi.org/JCDR/2012/3642:1937.

Ghanbari, A. and M.M.Attar (2014) Surface free energy characterization and adhesion performance of mild steel treated based on zirconium conversion coating: a comparative study. *Surface and Coatings Technology*, 246, 26–33.

Gorana, V.K., V.K.Jain, and G.K.Lal (2004) Experimental investigation into cutting forces and active grain density during abrasive flow machining. *International Journal of Machine Tools and Manufacture*, 44(2–3), 201–211.

Gov, K., O.Eyercioglu, and M.V.Cakir (2013) Hardness effects on abrasive flow machining. *Journal of Mechanical Engineering*, 59, 626–631.

Graf, W. (2015) Polish grinding of gears for higher transmission efficiency. *American Gear Manufacturers Association (AGMA) Fall Technical Meeting*, Detroit, MI, 15FTM21.

Gui, N., W.Xu, J.Tian, G.Rosengarten, M.Brandt, and M.Qian (2018) Fabrication and anisotropic wettability of titanium-coated microgrooves. *Journal of Applied Physics*, 123(9), 095306.

Guo, P., Y.Lu, K.F.Ehmann, and J.Cao (2014) Generation of hierarchical micro-structures for anisotropic wetting by elliptical vibration cutting. *CIRP Annals – Manufacturing Technology*, 63(1), 553 556.

Harkous, A., G.Colomines, E.Leroy, P.Mousseau, and R.Deterre (2016) The kinetic behavior of liquid silicone rubber: a comparison between thermal and rheological approaches based on gel point determination. *Reactive and Functional Polymers*, 101, 20–27.

Hashimoto, F., H.Yamaguchi, P.Krajnik, K.Wegener, R.Chaudhari, H.W.Hoffmeister, and F.Kuster (2016) Abrasive fine-finishing technology. *CIRP Annals – Manufacturing Technology*, 65(2), 597–620.

Hauslich, L.B., M.N.Sela, D.Steinberg, G.Rosen, and D.Kohavi (2013) The adhesion of oral bacteria to modified titanium surfaces: role of plasma proteins and electrostatic forces. *Clinical Oral Implants Research*, 24, 49–56.

Heisel, U. and J.Avroutine (2001) Process analysis for the evaluation of the surface formation and removal rate in lapping. *Annals of the CIRP*, 50(1), 229–232.

Hocevar, M., M.Jenko, M.Godec, and D.Drobne (2014) An overview of the influence of stainless-steel surface properties on bacterial adhesion. *Materials and Technology*, 48(48), 609–617.

Hull, J.B., D.O'Sullivan, A.J.Fletcher, S.A.Trengove, and J.Mackie (1992) Rheology of carrier media used in abrasive flow machining. *Key Engineering Materials Vols*, 72–74, 617–626.

Jain, R.K. and V.K.Jain (1999) Simulation of surface generated in abrasive flow machining process. *Robotics and Computer-Integrated Manufacturing*, 15, 403–412.

Jain, R.K. and V.K.Jain (2001) Specific energy and temperature determination in abrasive flow machining process. *International Journal of Machine Tools and Manufacture*, 41, 1689–1704.

Jain, R.K. and V.K.Jain (2004) Stochastic simulation of active grain density in abrasive flow machining. *Journal of Materials Processing Technology*, **152**(1), 17–22.

Jain, R.K., V.K.Jain, and P.M.Dixit (1999) Modeling of material removal and surface roughness in abrasive flow machining process. *International Journal of Machine Tools and Manufacture*, **39**(12), 1903–1923.

Jain, V.K. and S.G.Adsul (2000) Experimental investigations into abrasive flow machining (AFM). *International Journal of Machine Tools and Manufacture*, **40**, 1003–1021.

Jain, V.K., C.Ranganatha, and K.Muralidhar (2001) Evaluation of rheological properties of medium for AFM process. *Machining Science and Technology*, **5**(2), 151–170.

Jha, S. and V.K.Jain (2004) Design and development of the magnetorheological abrasive flow finishing (MRAFF) process. *International Journal of Machine Tools and Manufacture*, **44**, 1019–1029.

Jung, D., W.L.Wang, A.Knafl, T.J.Jacobs, S.J.Hu, and D.N.Assanis (2008) Experimental investigation of abrasive flow machining effects on injector nozzle geometries, engine performance, and emissions in a DI diesel engine. *International Journal of Automotive Technology*, **9**(1), 9–15.

Kang, C.-W. and F.-Z.Fang (2018) State of the art of bioimplants manufacturing: part I. *Advances in Manufacturing*, **6**(1), 20–40.

Kanthababu, M., M.S.Shunmugam, and M.Singaperumal (2009) Identification of significant parameters and appropriate levels in honing of cylinder liners. *International Journal of Machining and Machinability of Materials*, **5**(1), 80–96.

Kar, K.K., N.L.Ravikumar, P.B.Tailor, J.Ramkumar, and D.Sathiyamoorthy (2009a) Preferential media for abrasive flow machining. *Journal of Manufacturing Science and Engineering*, **131**(1), 011009.

Kar, K.K., N.L.Ravikumar, P.B.Tailor, J.Ramkumar, and D.Sathiyamoorthy (2009b) Performance evaluation and rheological characterization of newly developed butyl rubber based media for abrasive flow machining process. *Journal of Materials Processing Technology*, **209**(4), 2212–2221.

Katsikogianni, M. and Y.F.Missirlis (2004) Concise review of mechanisms of bacterial adhesion to biomaterials and of techniques used in estimating bacteria-material interactions. *European Cells and Materials*, **8**, 37–57.

Kavithaa, T.S. and N.Balashanmugam (2016) Nanometric surface finishing of typical industrial components by abrasive flow finishing. *The International Journal of Advanced Manufacturing Technology*, **85**, 2189–2196.

Kemaloglu, S., G.Ozkoc, and A.Aytac (2010) Properties of thermally conductive micro and nano size boron nitride reinforced silicon rubber composites. *ThermochimicaActa*, **499**(1–2), 40–47.

Kenda, J., J.Duhovnik, J.Tavcar, and J.Kopac (2014) Abrasive flow machining applied to plastic gear matrix polishing. *International Journal of Advanced Manufacturing Technology*, **71**(1–4), 141–151.

Kim, J.-D. and K.-D.Kim (2004) Deburring of burrs in spring collets by abrasive flow machining. *International Journal of Advanced Manufacturing Technology*, **24**, 469–473.

Koseki, H., A.Yonekura, T.Shida, I.Yoda, H.Horiuchi, Y.Morinaga, K.Yanagihara, H.Sakoda, M.Osaki, and M.Tomita (2014) Early staphylococcal biofilm formation on solid orthopaedic implant materials: in vitro study. *PLoS One*, **9**(10), 1–8.

Kubiak, K.J., M.C.T.Wilson, T.G.Mathia, and P.Carval (2011) Wettability versus roughness of engineering surfaces. *Wear*, **271**, 523–528.

Kumar, S., V.K.Jain, and A.Sidpara (2015) Nanofinishing of freeform surfaces (knee joint implant) by rotational-magnetorheological abrasive flow finishing (R-MRAFF) process. *Precision Engineering*, **42**, 165–178.

Laheurte, R., P.Darnis, N.Darbois, O.Cahuc, and J.Neauport (2012) Subsurface damage distribution characterization of ground surfaces using abbott–firestone curves. *Optics Express*, **20**(12), 13551.

Law, K.-Y. and H.Zhao (2016) *Surface Wetting: Characterization, Contact Angle, and Fundamentals*, Springer International Publishing, Switzerland.

Li, J., L.Yang, W.Liu, X.Zhang, and F.Sun (2014) Experimental research into technology of abrasive flow machining nonlinear tube runner. *Advances in Mechanical Engineering*, **6**, 752353.

Li, P., J.Xie, and Z.Deng (2015) Characterization of irregularly micro-structured surfaces related to their wetting properties. *Applied Surface Science*, **335**, 29–38.

Liang, Y., L.Shu, W.Natsu, and F.He (2015) Anisotropic wetting characteristics versus roughness on machined surfaces of hydrophilic and hydrophobic materials. *Applied Surface Science*, **331**, 41–49.

Loberg, J., I.Mattisson, S.Hansson, and E.Ahlberg (2010) Characterisation of titanium dental implants I: critical assessment of surface roughness parameters. *The Open Biomaterials Journal*, **2**, 18–35.

Loveless, T.R., R.E.Williams, and K.P.Rajurkar (1994) A study of the effects of abrasive-flow finishing on various machined surfaces. *Journal of Materials Processing Technology*, **47**(1–2), 133–151.

Lysaght, M.J. and J.A.O'Loughlin (2000) Demographic scope and economic magnitude of contemporary organ replacement therapies. *ASAIO Journal*, **46**(5), 515–521.

Mahajan, A. and S.S.Sidhu (2018) Surface modification of metallic biomaterials for enhanced functionality: a review. *Materials Technology*, **33**(2), 93–105.

Mali, H.S. and A.Manna (2010) Optimum selection of abrasive flow machining conditions during fine finishing of al/15 wt% SiC-MMC using taguchi method. *The International Journal of Advanced Manufacturing Technology*, **50**(9–12), 1013–1024.

Mattox, D.M. (2010) *Handbook of Physical Vapor Deposition (PVD) Processing*, 2nd Edition, William Andrew Imprint of Elsevier, Kidlington, London.

May, A., A.Agarwal, J.Lee, M.Lambert, C.K.Akkan, F.P.Nothdurft, and O.C.Aktas (2015) Laser induced anisotropic wetting on Ti-6Al-4V surfaces. *Materials Letter*, **138**, 21–24.

Menezes, P.L., S.V.Kailas, and M.R.Lovell (2013) *Fundamentals of Engineering Surfaces. Tribology for Scientists and Engineers*, Springer, New York.

Mezger, T.G. (2006) *The Rheology Handbook*, 2nd Edition, Vincentz Network, Hannover, Germany.

Moriarty, T.F., R.Kuehl, T.Coenye, W.-J.Metsemakers, M.Morgenstern, E.M.Schwarz, M.Riool, S.A.J.Zaat, N.Khana, and S.L.Kates (2016) Orthopaedic device-related infection: current and future interventions for improved prevention and treatment. *EFORT Open Reviews*, **1**(4), 89–99.

Mortimer, C.J., L.Burke, and C.J.Wright (2016) Microbial interactions with nanostructures and their importance for the development of electrospunnanofibrous materials used in regenerative medicine and filtration. *Journal of Microbial & Biochemical Technology*, **8**, 195–201.

Navarro, M., A.Michiardi, O.Castano, and J.A.Planell (2008) Biomaterials in orthopaedics. *Journal of the Royal Society Interface*, **5**(27), 1137–1158.

Nuraje, N., W.S.Khan, Y.Lei, M.Ceylan, and R.Asmatulu (2013) Superhydrophobicelectrospunnanofibers. *Journal of Materials Chemistry A*, **1**(6), 1929–1946.

Perry, W.B. and J.Stackhouse (1989) Gas turbine applications of abrasive flow machining. *Presented at the Gas Turbine and Aeroengine Congress and Exposition*, Toronto, Canada, June 4–8.

Petare, A.C. and N.K.Jain (2018) A critical review of past research and advances in abrasive flow finishing process. *The International Journal of Advanced Manufacturing Technology*, **97**(1–4), 741–782.

Petri, K.I., R.E.Billo, and B.Biclanda (1998) A neural network process model for abrasive flow machining operations. *Journal of Manufacturing Systems*, **17**(1), 52–64.

Pignon, F., A.Magnin, and J.Piau (1996) Thixotropic colloidal suspensions and flow curves with minimum: identification of flow regimes and rheometric consequences. *Journal of Rheology*, **40**(4), 573–587.

Quirynen, M. and C.M.L.Bollen (2005) The influence of surface roughness and surface-free energy on supra- and subgingival plaque formation in man. *Journal of Clinical Periodontology*, **22**(1), 1–14.

Raju, H.P., K.Narayanasamy, Y.G.Srinivasa, and R.Krishnamurthy (2003) Material response in extrusion honing. *Journal of Materials Science Letters*, **22**(5), 367–370.

Rasouli, R., A.Barhoum, and H.Uludag (2018) A review of nanostructured surfaces and materials for dental implants: surface coating, patterning and functionalization for improved performance. *Biomaterials Science*, **6**(6), 1312–1338. doi:https://doi.org/10.1039/c8bm00021b.

Rhoades, L. (1991) Abrasive flow machining: a case study. *Journal of Materials Processing Technology*, **28**, 107–116.

Ribeiro, M., F.J.Monteiro, and M.P.Ferraz (2012) Infection of orthopedic implants with emphasis on bacterial adhesion process and techniques used in studying bacterial-material interactions. *Biomatter*, **2**(4), 176–194.

Rohm, S., C.Knoflach, W.Nachbauer, M.Hasler, L.Kaserer, J.Van Putten, S.H.Unterberger, and R.Lackner (2016) Effect of different bearing ratios on the friction between ultrahigh molecular weight polyethylene ski bases and snow. *ACS Applied Materials and Interfaces*, **8**(19), 12552–12557.

Rulison, C. (1999) *So You Want to Measure Surface Energy?* KRUSS Technical Notes 306, Borsteler Chausses 85, Hamburg, Germany, 1–8.

Sankar, M.R., V.K.Jain, and J.Ramkumar (2009c) Experimental investigations into rotating workpiece abrasive flow finishing. *Wear*, **267**, 43–51.

Sankar, M.R., V.K.Jain, J.Ramkumar, and Y.M.Joshi (2011) Rheological characterization of styrene-butadiene based medium and its finishing performance using rotational abrasive flow finishing process. *International Journal of Machine Tools and Manufacture*, **51**(12), 947–957.

Sankar, M.R., S.Mondal, and J.Ramkumar (2009b) Experimental investigations and modeling of drill bit-guided abrasive flow finishing (DBG-AFF) process. *International Journal of Advanced Manufacturing Technology*, **42**(7–8), 678–688.

Sankar, M.R., J.Ramkumar, and V.K.Jain (2009a) Experimental investigation and mechanism of material removal in nano finishing of MMCS using abrasive flow finishing (AFF) process. *Wear*, **266**(7–8), 688–698.

Sharma, A.K., G.Venkatesh, S.Rajesha, and P.Kumar (2015) Experimental investigations into ultrasonic-assisted abrasive flow machining (UAAFM) process. *International Journal of Advanced Manufacturing Technology*, **80**, 477–493.

Shit, S.C. and P.Shah (2013) A review on silicone rubber. *National Academy Science Letters*, **36**(4), 355–365.

Sidpara, A.M. and V.K.Jain (2012) Nanofinishing of freeform surfaces of prosthetic knee joint implant. *ProcIMechE Part B: J Engineering Manufacture*, **226**(11), 1833–1846.

Singh, S. and H.S.Shan (2002) Development of magneto abrasive flow machining process. *International Journal of Machine Tools and Manufacture*, **42**(8), 953–959.

Singh, S., H.S.Shan, and P.Kumar (2008) Experimental studies on mechanism of material removal in abrasive flow machining process. *Materials and Manufacturing Processes*, **23**, 714–718.

Smith, G.T. (2002) *Industrial Metrology: Surface and Roundness*, 1st Edition, Springer-Verlag, London.

Song, F., H.Koo, and D.Ren (2015) Effects of material properties on bacterial adhesion and biofilm formation. *Journal of Dental Research*, **94**(8), 1027–1034.

Subramani, K., R.T.Mathew, and P.Pachauri (2018) *Titanium Surface Modification Techniques for Dental Implants-From Microscale to Nanoscale*, Second Edition, Elsevier Inc. doi:https://doi.org/10.1016/B978-0-12-812291-4.00006-6.

Sushil, M., K.Vinod, and K.Harmesh (2015) Experimental investigation and optimization of process parameters of Al/SiCMMCs finished by abrasive flow machining. *Materials and Manufacturing Processes*, **30**(7), 902–911.

Sykaras, N., A.M.Iacopino, V.A.Marker, R.G.Triplett, and R.D.Woody (2000) Implant materials, designs, and surface topographies: their effect on osseointegration. *A Literature Review Int J Oral Maxillofac Implants*, **2000**(15), 675–690.

Trengove, S.A. (1993) *Extrusion Honing Using Mixtures of Polyborosiloxane and Grit*, Ph.D. Thesis, Sheffield Hallam University, Sheffield, England.

Tseng, H.-C., J.-S.Wu, and R.-Y.Chang (2010) Linear viscoelasticity and thermorheological simplicity of n-hexadecane fluids under oscillatory shear via non-equilibrium molecular dynamics simulations. *Physical Chemistry Chemical Physics*, **12**(16), 4051.

Turger, A., J.Kohler, B.Denkena, T.A.Correa, C.Becher, and C.Hurschler (2013) Manufacturing conditioned roughness and wear of biomedical oxide ceramics for all-ceramic knee implants – Colorado school of mines. *BioMedical Engineering OnLine*, **12**(84), 1–17.

Tzeng, H.J., B.H.Yan, R.T.Hsu, and H.M.Chow (2007a) Finishing effect of abrasive flow machining on micro slit fabricated by wire-EDM. *International Journal of Advanced Manufacturing Technology*, **34**(7–8), 649–656.

Tzeng, H.J., B.H.Yan, R.T.Hsu, and Y.C.Lin (2007b) Self-modulating abrasive medium and its application to abrasive flow machining for finishing micro channel surfaces. *International Journal of Advanced Manufacturing Technology*, **32**(11–12), 1163–1169.

Uhlmann, E., M.Doits, and C.Schmiedel (2013) Development of a material model for visco-elastic abrasive medium in abrasive flow machining. *Procedia CIRP*, **8**, 351–356.

Uhlmann, E., V.Mihotovic, and A.Coenen (2009) Modelling the abrasive flow machining process on advanced ceramic materials. *Journal of Materials Processing Technology*, **209**, 6062–6066.

van-Oss, C.J., M.K.Chaudhury, and R.J.Good (1988) Interfacial lifshitz-van der waals and polar interactions in macroscopic systems. *Chemical Reviews*, **88**(6), 927–941.

Walia, R.S., H.S.Shan, and P.Kumar (2006a) Abrasive flow machining with additional centrifugal force applied to the media. *Machining Science and Technology*, **10**(3), 341–354.

Walia, R.S., H.S.Shan, and P.Kumar (2006b) Parametric optimization of centrifugal force-assisted abrasive flow machining (CFAAFM) by the taguchi method. *Materials and Manufacturing Processes*, **21**(4), 375–382.

Wang, A.C., K.C.Cheng, K.Y.Chen, and Y.C.Lin (2014) Enhancing the surface precision for the helical passageways in abrasive flow machining. *Materials and Manufacturing Processes*, **29**(2), 153–159.

Wang, A.C., C.H.Liu, K.Z.Liang, and S.H.Pai (2007) Study of the rheological properties and the finishing behavior of abrasive gels in abrasive flow machining. *Journal of Mechanical Science and Technology*, **21**(10), 1593–1598.

Wang, A.C. and S.H.Weng (2007) Developing the polymer abrasive gels in AFM processs. *Journal of Materials Processing Technology*, **192–193**, 486–490.

Wassmann, T., S.Kreis, M.Behr, and R.Buergers (2017) The influence of surface texture and wettability on initial bacterial adhesion on titanium and zirconium oxide dental implants. *International Journal of Implant Dentistry*, **3**(32), 1–11.

Wenzel, R. (1936) Resistance of solid surfaces. *Journal of Industrial and Engineering Chemistry*, **28**(8), 988–994.

Whitehouse, D.J. (2011) *Handbook of Surface and Nanometrology*, 2nd Edition, CRC Press, Taylor & Francis Group, Boca Raton, FL, London and New York.

Williams, R.E. (1993) *Investigation of the Abrasive Flow Machining Process and Development of a Monitoring Strategy Using Acoustic Emission*, Ph.D Thesis, University of Nebraska, Lincoln, NE.

Williams, R.E. (1998) Acoustic emission characteristics of abrasive flow machining. *Journal of Manufacturing Science and Engineering*, **120**(2), 264–271.

Williams, R.E. and K.P.Rajurkar (1992) Stochastic modeling and analysis of abrasive flow machining. *Journal of Manufacturing Science and Engineering*, **114**(1), 74–81.

Winter, H.H. (1987) Can the gel point of a cross-linking polymer be detected by the G′ -G″ crossover? *Polymer Engineering and Science*, **27**(22), 1698–1702.

Wu, S. (1982) *Polymer Interface and Adhesion*, 1st Edition, CRC Press, New York.

Xu, Y.C., K.H.Zhang, S.Lu, and Z.Q.Liu (2013) Experimental investigations into abrasive flow machining of helical gear. *Key Engineering Materials*, **546**, 65–69.

Yan, Y., E.Chibowski, and A.Szczes (2017) Surface properties of Ti-6Al-4V alloy part I: surface roughness and apparent surface free energy. *Materials Science and Engineering: C*, **70**, 207–215.

Yin, L., K.Ramesh, S.Wan, X.D.Liu, H.Huang, and Y.C.Liu (2004) Abrasive flow polishing of micro bores. *Materials and Manufacturing Processes*, **19**(2), 187–207.

Yoda, I., H.Koseki, M.Tomita, T.Shida, H.Horiuchi, H.Sakoda, and M.Osaki (2014) Effect of surface roughness of biomaterials on staphylococcus epidermidis adhesion. *BMC Microbiology*, **14**(1), 1–7.

Zhou, W., S.Qi, H.Zhao, and N.Liu (2007) Thermally conductive silicone rubber reinforced with boron nitride particle. *Polymer Composites*, **28**(1), 23–28.

Index

4/3 solenoid operated direction control valve, specification of, 65

Printed in the United States
by Baker & Taylor Publisher Services